LONDON MATHEMATICAL SOCIETY LECTURE NOTE SERIES

Managing Editor: Professor J.W.S. Cassels, Department of Pure Mathematics and Mathematical Statistics, 16 Mill Lane, Cambridge CB2 1SB, England

The books in the series listed below are available from booksellers, or, in case of difficulty, from Cambridge University Press.

4 Algebraic topology, J.F.ADAMS
5 Commutative algebra, J.T.KNIGHT
8 Integration and harmonic analysis on compact groups, R.E.EDWARDS
11 New developments in topology, G.SEGAL (ed)
12 Symposium on complex analysis, J.CLUNIE & W.K.HAYMAN (eds)
13 Combinatorics, T.P.McDONOUGH & V.C.MAVRON (eds)
16 Topics in finite groups, T.M.GAGEN
17 Differential germs and catastrophes, Th.BROCKER & L.LANDER
18 A geometric approach to homology theory, S.BUONCRISTIANO, C.P.ROURKE & B.J.SANDERSON
20 Sheaf theory, B.R.TENNISON
21 Automatic continuity of linear operators, A.M.SINCLAIR
23 Parallelisms of complete designs, P.J.CAMERON
24 The topology of Stiefel manifolds, I.M.JAMES
25 Lie groups and compact groups, J.F.PRICE
26 Transformation groups, C.KOSNIOWSKI (ed)
27 Skew field constructions, P.M.COHN
29 Pontryagin duality and the structure of LCA groups, S.A.MORRIS
30 Interaction models, N.L.BIGGS
31 Continuous crossed products and type III von Neumann algebras,A.VAN DAELE
32 Uniform algebras and Jensen measures, T.W.GAMELIN
34 Representation theory of Lie groups, M.F. ATIYAH et al.
35 Trace ideals and their applications, B.SIMON
36 Homological group theory, C.T.C.WALL (ed)
37 Partially ordered rings and semi-algebraic geometry, G.W.BRUMFIEL
38 Surveys in combinatorics, B.BOLLOBAS (ed)
39 Affine sets and affine groups, D.G.NORTHCOTT
40 Introduction to Hp spaces, P.J.KOOSIS
41 Theory and applications of Hopf bifurcation, B.D.HASSARD, N.D.KAZARINOFF & Y-H.WAN
42 Topics in the theory of group presentations, D.L.JOHNSON
43 Graphs, codes and designs, P.J.CAMERON & J.H.VAN LINT
44 Z/2-homotopy theory, M.C.CRABB
45 Recursion theory: its generalisations and applications, F.R.DRAKE & S.S.WAINER (eds)
46 p-adic analysis: a short course on recent work, N.KOBLITZ
47 Coding the Universe, A.BELLER, R.JENSEN & P.WELCH
48 Low-dimensional topology, R.BROWN & T.L.THICKSTUN (eds)
49 Finite geometries and designs,P.CAMERON, J.W.P.HIRSCHFELD & D.R.HUGHES (eds)
50 Commutator calculus and groups of homotopy classes, H.J.BAUES
51 Synthetic differential geometry, A.KOCK
52 Combinatorics, H.N.V.TEMPERLEY (ed)
54 Markov process and related prob[illegible] DYNKIN
55 Ordered permutation groups, A.[illegible]
56 Journêes arithmêtiques, J.V.AR[illegible]
57 Techniques of geometric topolo[illegible]
58 Singularities of smooth functi[illegible]
59 Applicable differential geomet[illegible]
60 Integrable systems, S.P.NOVIK[illegible]
61 The core model, A.DODD
62 Economics for mathematicians, J.W.S.CASSELS
63 Continuous semigroups in Banach algebras, A.M.SINCLAIR

64 Basic concepts of enriched category theory, G.M.KELLY
65 Several complex variables and complex manifolds I, M.J.FIELD
66 Several complex variables and complex manifolds II, M.J.FIELD
67 Classification problems in ergodic theory, W.PARRY & S.TUNCEL
68 Complex algebraic surfaces, A.BEAUVILLE
69 Representation theory, I.M.GELFAND et al.
70 Stochastic differential equations on manifolds, K.D.ELWORTHY
71 Groups - St Andrews 1981, C.M.CAMPBELL & E.F.ROBERTSON (eds)
72 Commutative algebra: Durham 1981, R.Y.SHARP (ed)
73 Riemann surfaces: a view towards several complex variables,A.T.HUCKLEBERRY
74 Symmetric designs: an algebraic approach, E.S.LANDER
75 New geometric splittings of classical knots, L.SIEBENMANN & F.BONAHON
76 Linear differential operators, H.O.CORDES
77 Isolated singular points on complete intersections, E.J.N.LOOIJENGA
78 A primer on Riemann surfaces, A.F.BEARDON
79 Probability, statistics and analysis, J.F.C.KINGMAN & G.E.H.REUTER (eds)
80 Introduction to the representation theory of compact and locally compact groups, A.ROBERT
81 Skew fields, P.K.DRAXL
82 Surveys in combinatorics, E.K.LLOYD (ed)
83 Homogeneous structures on Riemannian manifolds, F.TRICERRI & L.VANHECKE
84 Finite group algebras and their modules, P.LANDROCK
85 Solitons, P.G.DRAZIN
86 Topological topics, I.M.JAMES (ed)
87 Surveys in set theory, A.R.D.MATHIAS (ed)
88 FPF ring theory, C.FAITH & S.PAGE
89 An F-space sampler, N.J.KALTON, N.T.PECK & J.W.ROBERTS
90 Polytopes and symmetry, S.A.ROBERTSON
91 Classgroups of group rings, M.J.TAYLOR
92 Representation of rings over skew fields, A.H.SCHOFIELD
93 Aspects of topology, I.M.JAMES & E.H.KRONHEIMER (eds)
94 Representations of general linear groups, G.D.JAMES
95 Low-dimensional topology 1982, R.A.FENN (ed)
96 Diophantine equations over function fields, R.C.MASON
97 Varieties of constructive mathematics, D.S.BRIDGES & F.RICHMAN
98 Localization in Noetherian rings, A.V.JATEGAONKAR
99 Methods of differential geometry in algebraic topology, M.KAROUBI & C.LERUSTE
100 Stopping time techniques for analysts and probabilists, L.EGGHE
101 Groups and geometry, ROGER C.LYNDON
102 Topology of the automorphism group of a free group, S.M.GERSTEN
103 Surveys in combinatorics 1985, I.ANDERSEN (ed)
104 Elliptical structures on 3-manifolds, C.B.THOMAS
105 A local spectral theory for closed operators, I.ERDELYI & WANG SHENGWANG
106 Syzygies, E.G.EVANS & P.GRIFFITH
107 Compactification of Siegel moduli schemes, C-L.CHAI
108 Some topics in graph theory, H.P.YAP
109 Diophantine analysis, J.LOXTON & A.VAN DER POORTEN (eds)
110 An introduction to surreal numbers, H.GONSHOR

London Mathematical Society Lecture Note Series. 110

An Introduction to the Theory of Surreal Numbers

HARRY GONSHOR
Rutgers University

CAMBRIDGE UNIVERSITY PRESS
Cambridge
London New York New Rochelle
Melbourne Sydney

Published by the Press Syndicate of the University of Cambridge
The Pitt Building, Trumpington Street, Cambridge CB2 1RP
32 East 57th Street, New York, NY 10022, USA
10, Stamford Road, Oakleigh, Melbourne 3166, Australia

First published 1986

Printed in Great Britain at the University Press, Cambridge

Library of Congress cataloging in publication data

Gonshor, H.
An introduction to the theory of surreal numbers (London Mathematical Society lecture note series; 110)
Bibliography: p.
1. Numbers, Theory of. I. Title. II. Series
QA241.G63 1986 512'.7 86-9668

British Library cataloguing in publication data

Gonshor, H.
An introduction to the theory of surreal numbers - (London Mathematical Society lecture note series, ISSN 0076-0052; 110
1. Numbers, Theory of, 2. Logic
I. Title II. Series
512'.7 QA241

ISBN 0 521 31205 1

CONTENTS

PREFACE

The aim of this book is to give a systematic introduction to the theory of surreal numbers based on foundations that are familiar to most mathematicians. I feel that the surreal numbers form an exciting system which deserves to be better known and that therefore an exposition like this one is needed at present. The subject is in such a pioneering state that it appears that there are many results just on the verge of being discovered and even concepts that still are waiting to be defined.

One might claim that one should wait till the theory of surreal numbers is more fully established before publishing a book on this subject. Such a comment reminds me of the classic joke about the person who is afraid of drowning and has vowed never to step into water until he has learned how to swim. In fact, the time is ripe for such a book and furthermore the book itself should contribute to developing the subject with the help of creative readers.

The subject has suffered so far from isolation with pockets of people in scattered parts of the world working on those facets of the subject that interest them. I hope that this book will play a role in eliminating this isolation and bringing together the mathematicians interested in surreal numbers.

The book is thus a reflection of my own personal interest. For example, Martin Kruskal has developed the theory of exponentiation from a somewhat different point of view and carried it in different directions from the presentation in this book. Also, I recently received correspondence from Norman Alling who has recently done work on a facet of the theory of surreal numbers not discussed in the book. With greater communication all this and more could play a role in a future edition.

The basic material is found in chapters 2 through 5. The later chapters are more original and more specialized. Although room for future improvement exists everywhere, chapters 7 and 8 are in an especially pioneering position: this is where the greatest opportunity seems to exist for knowledgeable readers to obtain new results.

ACKNOWLEDGEMENTS

I would like to thank the following people for their help in connection with the manuscript.

First, there is Professor Larry Corwin who took time from his very busy schedule to do a great deal of proofreading. He is responsible for many improvements in the exposition throughout the manuscript. On the other hand, I take full responsibility for any faults in the exposition which still remain. Professor Joe D'Atri and Jim Maloney, a graduate student, have also helped with some proofreading. Also, I should mention Professor Barbara Osofsky who much earlier had read a preliminary draft of chapters one and two and made many valuable suggestions.

Finally, I mention the contribution of two secretaries of the Rutgers Mathematics Department. Mary Anne Jablonski, the co-ordinator, took care of numerous technical details, and Adelaide Boulle did an excellent job of turning my handwritten draft into typescript.

1 INTRODUCTION

The surreal numbers were discovered by J.H. Conway. He was mainly interested in games for which he built up a formalism for generalizing the classical theory of impartial games. Numbers were obtained as special cases of games. Donald E. Knuth began a study of these numbers in a little book [2] in the form of a novel in which the characters are trying to use their creative talents to discover proofs. Conway goes into much more depth in his classic book <u>On Numbers and Games</u> [1].

I was introduced to this subject in a talk by M.D. Kruskal at the A.M.S. meeting in St. Louis in January 1977. Since then I have developed the subject from a somewhat different foundation from Conway, and carried it further in several directions. I define the surreal numbers as objects which are rather concrete to most mathematicians, as compared to Conway's, which are equivalence classes of inductively defined objects.

The surreal numbers form a proper class which contains the real numbers and the ordinals among other things. For example, in this system $\omega-1$, $\sqrt{\omega}$, etc. make sense and, in fact, arise naturally. I believe that this system is of sufficient interest to be worthy of being placed alongside the other systems that are being studied by mathematicians. First, as we shall see, we obtain a nice way of building up the real number system. Instead of being compelled to create new entities at each stage and make new definitions, we have unified definitions at the beginning and obtain the reals as a subsystem of what we already have. Secondly, and more important than obtaining a new way of building up a familiar set such as the real numbers, is the enrichment of mathematics by the inclusion of a new structure with interesting properties.

In fact, it is because the system seems to be so natural to the author that the first sentence contains the word "discovered" rather than "constructed" or "created." Thus the fact that the system was discovered so recently is somewhat surprising. Be that as it may, the pioneering nature of the subject gives any potential reader the opportunity of getting in on the ground floor. That is, there are practically no prerequisites for reading this book other than a little mathematical maturity. Thus the reader has the opportunity which is all too rare nowadays of getting to the surface and tackling interesting original problems without having to learn a huge amount of material in advance.

The only prerequisite worthy of mention is a minimal intuitive knowledge of ordinals, for example familiarity with the distinction between non-limit and limit ordinals. For a fuller understanding it is useful to be familiar with the basic operations of addition, multiplication, and exponentiation.

The results and some of the proofs in the earlier chapters are essentially the same as those in [1] but the theory begins with a different foundation. The later chapters tend to be more original. The ideas in Chapters 6 and 7 are new as far as I know. [1] contains several remarks related to chapter 9 where the ideas are studied in detail. Part of the material in chapter 10 was done independently by Kruskal. At present, his work is unpublished. I would like to give credit to Kruskal for pointing out to me that exponentiation can be defined in a natural way for the surreal numbers. Using his hints I developed the theory independently. Although naturally there is an overlap at the beginning, it appears from private conversations that Kruskal did not pursue the topics in sections C and D.

2 DEFINITION AND FUNDAMENTAL EXISTENCE THEOREM

A DEFINITION

Definition. A surreal number is a function from an initial segment of the ordinals into the set {+,-}, i.e. informally, an ordinal sequence consisting of pluses and minuses which terminate. The empty sequence is included as a possibility.

Examples. One example is the function f defined as f(0) = +, f(1) = - and f(2) = + which is informally written as (+-+). An example of infinite length is the sequence of ω pluses followed by ω minuses.

The length $\ell(a)$ of a surreal number is the least ordinal α for which it is undefined. (Since an ordinal is the set of all its predecessors this is the same as the domain of a, but I prefer to avoid this point of view.) An initial segment of a is a surreal number b such that $\ell(b) \leq \ell(a)$ and $b(\alpha) = a(\alpha)$ for all α where $b(\alpha)$ is defined. The tail of b in a is the surreal number c of length $\ell(a)-\ell(b)$ satisfying $c(\alpha) = a[\ell(b)+\alpha]$. Informally, this is the sequence obtained from a by chopping off b from the beginning. a may be regarded as the juxtaposition of b and c written bc.

For stylistic reasons I shall occasionally say that $a(\alpha) = 0$ if a is undefined at α. This should be regarded as an abuse of notation since we do not want the domain of a to be the proper class of all ordinals.

Definition. If a and b are surreal numbers we define an order as follows:

a < b if $a(\alpha) < b(\alpha)$ where α is the first place where a and b differ, with the convention that - < 0 < +, e.g. (+-) < (+) < (++).

It is clear that this is a linear order. In fact, this is essentially a lexicographical order.

B FUNDAMENTAL EXISTENCE THEOREM

Theorem 2.1. Let F and G be two sets of surreal numbers such that $a \in F$ and $b \in G \Rightarrow a < b$. Then there exists a unique c of minimal length such that $a \in F \Rightarrow a < c$ and $b \in G \Rightarrow c < b$. Furthermore c is an initial segment of any surreal number strictly between F and G. (Note that F or G may be empty.)

Note. Henceforth I use the natural convention that if F and G are sets then "$F < G$" means "$a \in F$ and $b \in G \Rightarrow a < b$," "$F < c$" means "$a \in F \Rightarrow a < c$" and "$c < G$" means "$b \in G \Rightarrow c < b$." Thus we may write the hypothesis as $F < G$.

Example. Let F consist of all finite sequences of pluses and G be the unit set whose only member is the sequence of ω pluses. Then $F < G$. It is trivial to verify directly that c consists of ω pluses followed by a minus, i.e., $F < c < G$ and that any sequence d satisfying $F < d < G$ begins with c.

This theorem makes an alternative approach to the one in [1] possible. In [1] the author regards pairs (F,G) as abstract objects where the elements in F and G have been previously defined by the same method, as pairs of sets. (It is possible to start this induction by letting F and G both be the null set.) Since different pairs can give rise to the same number, the author needs an inductively defined equivalence relation. Theorem 2.1 gives us a definite number corresponding to the pair (F,G) so that we dispense with abstract pairs.

Proof. Clearly, it suffices to prove the initial segment property.

Case 1. If F and G are empty, then clearly the empty sequence works.

Case 2. G is empty but F is nonempty.

Let α be the least ordinal such that there does not exist $a \in F$ such that $a(\beta) = +$ for all $\beta < \alpha$. Thus α cannot equal zero, since any a vacuously satisfies the condition $a(\beta) = +$ for all $\beta < 0$.

<u>Subcase 1</u>. α is a limit ordinal. I claim that the desired c is the sequence of α pluses, i.e., $\ell(c) = \alpha$ and $c(\beta) = +$ if $\beta < \alpha$.

Since, by choice of α, no element a of F exists such that $a(\beta) = +$ for all $\beta < \alpha$, every element of F is less than c.

Now let d be any surreal number such that $F < d$.

Suppose $\gamma < \alpha$. Then $\gamma + 1 < \alpha$, since α is a limit ordinal. Hence, by choice of α, there exists $a \in F$ such that $a(\beta) = +$ for all $\beta < \gamma+1$, i.e. $\beta \leq \gamma$. Since $a < d$, $d(\beta) = +$ for all $\beta \leq \gamma$. In particular, $d(\gamma) = +$. Thus c is an initial segment of d.

<u>Subcase 2</u>. α is a non-limit ordinal, $\gamma + 1$. In this case there exists an $a \in F$ such that $a(\beta) = +$ for all $\beta < \gamma$ but there is no $a \in F$ such that $a(\beta) = +$ for all $\beta \leq \gamma$. Hence any $a \in F$ satisfying: $(\beta < \gamma \Rightarrow a(\beta) = +)$ must satisfy: $(a(\gamma) = -$ or $0)$. If all such a satisfy $a(\gamma) = -$ then it is easy to see that the sequence of γ pluses works for c. If there exist such an $a \in F$ such that $a(\gamma) = 0$, i.e. the sequence of γ pluses belongs to F, then the sequence of $(\gamma+1)$ pluses works for c.

<u>Case 3</u>. F is empty but G is nonempty. This case is similar to Case 2.

<u>Case 4</u>. Both F and G are nonempty.

Let α be the least ordinal such that there do not exist $a \in F$, $b \in G$ such that $a(\beta) = b(\beta)$ for all $\beta < \alpha$. Again $\alpha \neq 0$.

<u>Subcase 1</u>. α is a limit ordinal. Suppose $\gamma < \alpha$; then $\gamma+1 < \alpha$. Hence there exist $a \in F$, $b \in G$ such that $a(\beta) = b(\beta)$ for all $\beta \leq \gamma$. The value $a(\gamma)$ is well-defined in the following sense. If (a_1,b_1) is another pair satisfying the above properties then $a(\beta) = a_1(\beta)$ for all $\beta \leq \gamma$. Otherwise, suppose $\delta \leq \gamma$ is the least ordinal for which $a(\beta) \neq a_1(\beta)$. Without loss of generality assume $a(\delta) < a_1(\delta)$. Then by the lexicographical order $b < a_1$, which is a contradiction since $b \in G$ and $a_1 \in F$. Thus there exists a sequence of length α, such that for all $\gamma < \alpha$ there exist $a \in F$ and $b \in G$ such that $a(\beta) = d(\beta) = b(\beta)$ for $\beta \leq \gamma$.

By hypothesis on α, d cannot be an initial segment of an element in F as well as an element in G. Furthermore, an element of F which does not have d as an initial segment must be less than d. (Otherwise we obtain the same contradiction, as before.) Similarly an element of G which does not have d as an initial segment must be greater than d.

It follows that if d is neither an initial segment of an element of F nor an initial segment of an element of G then d works.

Now suppose F has elements with initial segment d. Then G does not have such elements. Let F' be the set of tails with respect to d of all such elements in F. Apply case 2 to F' and ϕ to obtain d'. Then the juxtaposition dd' works.

First, as before the required inequality is satisfied with respect to all elements in F or G which do not begin with d. Since $F' < d'$ it follows from the lexicographical order that dd' is larger than all elements in F beginning with d.

On the other hand, let e be any element satisfying $F < e < G$. Recall that for all $\gamma < \alpha$ there exist $a \in F$ and $b \in G$ such that $a(\beta) = d(\beta) = b(\beta)$ for $\beta \leq \gamma$. This implies by the lexicographical order that $e(\beta) = d(\beta)$ for $\beta < \alpha$. Thus d is an initial segment of e. Again using the lexicographical order the tail e must satisfy $F' < e'$. Hence d' is an initial segment of e'. Therefore dd' is an initial segment of e.

A similar argument applies if G has elements with initial segment d.

<u>Subcase 2</u>. α is a non-limit ordinal $\gamma+1$. Then there exist $a \in F$, $b \in G$ such that $a(\beta) = b(\beta)$ for all $\beta < \gamma$ but there do not exist $a \in F$, $b \in G$ which agree for all $\beta \leq \gamma$. As before, the values $a(\beta)$ are well-defined, and we obtain a sequence d of length γ. Again, as before, any element in F which does not have d as an initial segment must be less than d and an element in G which does not have d as an initial segment must be greater than d.

Let F' be the set of tails with respect to d of elements in F which begin with d and similarly for G'. Then as stated previously, there do not exist $a \in F'$, $b \in G'$ such that $a(0) = b(0)$. [Note that in contrast to subcase 1, F' and G' are non-empty although one of these sets might contain the empty sequence as its only element.] Since $F' < G'$, it follows that $a(0) < b(0)$ for all $a \in F'$, $b \in G'$.

Now suppose $d \in F$ and $d \in G$. This means that neither F' nor G' contains the empty sequence, i.e. $a(0)$ and $b(0)$ are never undefined. Since $a(0) < b(0)$, we obtain: $a(0) = -$ and $b(0) = +$. It is then clear that d works.

Since F and G are disjoint, d belongs to at most one of F and G. Suppose that $d \in G$. A similar argument will apply if $d \in F$. Then every a in F' satisfies $a(0) = -$. Let F'' be the set of tails of F' with respect to this $-$. (Such an iterated tail is, clearly, the tail with respect to the sequence $(d-.)$ Apply case 2 to F'' and ϕ to obtain d'. Then the juxtaposition $c = d-d'$ works. We already know that c satisfies the required inequality with respect to those elements that do not begin with d. Since no $b \in G'$ has $b(0) = -$, this takes care of all of G. The choice of d' takes care of all elements in F beginning with d (the next term of which is necessarily $-$). On the other hand, any element e satisfying $F < e < G$ must begin with d. Since $d \in G$, the next term must be $-$. By choice of d', it must be an initial segment of the tail of e with respect to $d-$, i.e. e must begin with $d-d'$.

This completes the proof.

Definition. $F|G$ is the unique c of minimal length such that $F < c < G$.

Remark. A slightly easier but less constructive proof is possible. First one extracts what is needed from the above proof to obtain an element c such that $F < c < G$. Although this is all that is required for the conclusion, the proof does not simplify tremendously. Nevertheless, it simplifies slightly since there is no concern about the initial segment property. Once a c is obtained, the well-ordering principle gives us a c of minimal length. At this stage it is useful to have a definition.

Definition. The common initial segment of a and b where $a \neq b$ is the element c whose length is the least α such that $a(\alpha) \neq b(\alpha)$ and such that $c(\beta) = a(\beta) = b(\beta)$ for $\beta < \alpha$. If $a = b$ then $c = a = b$.

If one of a or b is an initial segment of the other, then c is the shorter element. If neither is an initial segment of the other, then either $a(\gamma) = +$ and $b(\gamma) = -$ or $a(\gamma) = -$ and $b(\gamma) = +$. In either case c is strictly between a and b.

Now let c satisfy $F < c < G$ and be of minimal length. Suppose $F < d < G$. Let e be the common initial segment of c and d. Then $F < e < G$. Since c has minimal length and e is an initial segment of c, $e = c$. Hence $c = e$ is an initial segment of d.

C ORDER PROPERTIES

Theorem 2.2. If $G = \phi$ then $F|G$ consists solely of pluses.

Proof. This follows immediately from the construction in the proof of theorem 2.1. It can also be seen trivially as follows. Suppose c has minuses. Let d be the initial segment of c of length γ where γ is the least ordinal at which c has the value plus. Then clearly $F < d$ and d has shorter length than c. This contradicts the minimality of the length of c.

Theorem 2.2a. If $F = \phi$ then $F|G$ consists solely of minuses.

Proof. Similar to the above.

Note that the empty sequence consists solely of pluses and solely of minuses!

Theorem 2.3. $\ell(F|G) \leq$ the least α such that $\forall a[a \in F \cup G \Rightarrow \ell(a) < \alpha]$.

This is trivial because of the lexicographical order, since otherwise the initial segment b of $F|G$ of length α would also satisfy $F < b < G$ contradicting the minimality of $F|G$.

Note that $<$ cannot be replaced by $\leq$. For example, if $F = \{(+)\}$ and $G = \{(++)\}$, then $F|G = (++-)$. The result also follows from the construction in the proof of theorem 2.1. In fact, the construction gives the more detailed information that every proper initial segment of $F|G$ is an initial segment of an element of $F \cup G$. (An initial segment b of a is proper if $b \neq a$).

Theorem 2.3 has a kind of converse.

Theorem 2.4. Any a of length α can be expressed in the form $F|G$ where all elements of $F \cup G$ have length less than α.

Proof. Let $F = \{b: \ell(b) < \alpha$ and $b < a\}$ and $G = \{b: \ell(b) < \alpha$ and $b > a\}$. Then $F < a < G$ and every element of length less than α is, by definition, in F or G so that a satisfies the minimum length condition. Note that the argument is valid even if a is the empty sequence.

The last result is a step in the way of showing the connection between what is done here and the spirit of [1], since the result says that every element can be expressed in terms of elements of smaller length, thus every element can be obtained inductively by the methods of [1]. The next theorem shows that the ordering in [1] is equivalent to the one used here.

Theorem 2.5. Suppose $F|G = c$ and $F'|G' = d$. Then $c \le d$ iff $c < G'$ and $F < d$.

Proof. We know that $F < c < G$ and $F' < d < G'$. Suppose $c \le d$; then $c \le d < G'$ and $F < c \le d$. For the converse, assume $c < G'$ and $F < d$. We show that $d < c$ leads to a contradiction. This assumption yields $F < d < c < G$. Hence c is an initial segment of d. Also $F' < d < c < G'$ so d is an initial segment of c. Hence $c = d$ which contradicts $d < c$.

This last result is of minor interest for our purpose. Its main interest is that together with theorems 2.1 and 2.4 it shows that we are dealing with essentially the same objects as in [1] although here they are concretely defined. Since the present work is self-contained this is not of urgent importance, although it is worthy of noting.

Of fundamental importance here will be what I call the "cofinality theorems." They are analogous to classical results such as: In the ε,δ definition of a limit, it suffices to consider rational ε; and in the definition of a direct limit of objects with respect to a directed set, a cofinal subset gives rise to an isomorphic object.

Definition. (F',G') is cofinal in (F,G) if $(\forall a \varepsilon F)(\exists b \varepsilon F')(b \ge a) \wedge (\forall a \varepsilon G)(\exists b \varepsilon G')(b \le a)$.

It is clear that (F,G) is cofinal in (F,G), and that (F'',G'') cofinal in (F',G') and (F',G') cofinal in (F,G) implies (F'',G'') cofinal in (F,G). Also if $F \subset F'$ and $G \subset G'$ then (F',G') is cofinal in (F,G).

The following theorems are important although they are trivial to prove.

Theorem 2.6 (the cofinality theorem). Suppose $F|G = a$, $F' < a < G'$, and (F',G') is cofinal in (F,G); then $F'|G' = a$.

Proof. Suppose $\ell(b) < \ell(a)$ and $F' < b < G'$. It follows immediately from cofinality that $F < b < G$, contradicting the minimality of $\ell(a)$. Hence $a = F'|G'$.

Theorem 2.7 (cofinality theorem b). Suppose (F,G) and (F',G') are mutually cofinal in each other. Then $F|G = F'|G'$.

Note that it is enough to assume that $F|G$ has meaning since $F < G \Rightarrow F' < G'$.

Proof. $\{x:F<x<G\} = \{x:F'<x<G'\}$. Hence the element of minimal length is the same.

Although the two above theorems are closely related they are not quite the same. Theorem 2.6 will be especially useful in the sequel. I emphasize that in spite of the simplicity of the proof it is more convenient to quote the term "cofinality" than to repeat the trivial argument every time it is used. Also it is convenient often with abuse of notation to say that H' is cofinal in H. However, this is unambiguous only if it is clearly understood whether H and H' appear on the left or right, i.e. we must consider whether we are comparing (H,G) with (H',G') or (F,H) with (F',H'). This is usually clear from the context.

Cofinality will be used to sharpen theorem 2.4 to obtain the canonical representation of a as $F|G$. Of course, the representation in theorem 2.4 itself may be regarded as the "canonical" representation. The choice is simply a matter of taste.

Theorem 2.8. Let a be a surreal number. Suppose that $F' = \{b: b < a$

and b is an initial segment of a} and $G' = \{b: b > a$ and b is an initial segment of a}. Then $a = F'|G'$. (In the sequel $F'|G'$ will be called the canonical representation of a.)

Proof. We first use the representation in theorem 2.4. Then $F' \subset F$ and $G' \subset G$. Since it is clear that $F' < a < G'$, it suffices by theorem 2.6 to show that (F',G') is cofinal in (F,G). Let $b \in F$. Then $\ell(b) < \ell(a)$. Suppose c is the common initial segment of a and b. Then $b \leq c < a$. Hence $c \in F'$. A similar argument shows that G' is cofinal in G.

The above representation is especially succinct. It is easy to see that F' is the set of all initial segments of a of length β for those β such that $a(\beta) = +$ and similarly G' is the set of all initial segments of a of length β for those β such that $a(\beta) = -$. Thus the elements of F' and G' are naturally parametrized by ordinals. Furthermore, the elements of F form an increasing function of β and the elements of G form a decreasing function of β. Thus by a further use of the cofinality theorem we may restrict F' or G' to initial segments of length γ where the set of γ is cofinal in the set of β. For example, let a = (++-+--+). Then F' = {( ),(+),(++-),(++-+--)} and G' = {(++),(++-+),(++-+-)}. To avoid confusion it is important to recall that the ordinals begin with 0. So, e.g., a(3) = +. Hence the initial segment of length 3 = a(0),a(1),a(2)] = ++-. In other words, this terminates just before a(3) = + so that it really belongs to F'. Note the way F' and G' get closer and closer to a in a manner analogous to that of partial sums of an alternating series approximating its sum. However, the analogy is limited by the possibility of having many alike signs in a row; e.g., in the extreme case of all pluses, there are no approximations from above. As an application of the last remark on cofinal sets of ordinals we also have a = {(+),(++-+--)}|{(++),(++-+-)} or even more simply as {(++-+--)}|{(++-+-)}, since any subset containing the largest ordinal is cofinal in a finite set of ordinals.

In view of the above it seems natural to use F' and G' for the canonical representation of a. F and G on the other hand appear to contain lots of extra "garbage."

Finally, we need a result which may be regarded as a partial converse to the cofinality theorem. First, it is unreasonable to expect a true converse; in fact, it is surprising at first that any kind of converse is possible. If $a = F|G$ choose b so that $F < b < a$. Such b exists by theorem 2.1. By the cofinality theorem $F \cup \{b\}|G = a$. However, F is not cofinal in $F \cup \{b\}$ by choice of b.

Theorem 2.9 (the inverse cofinality theorem). Let $a = F|G$ be the canonical representation of a. Also let $a = F'|G'$ be an arbitrary representation. Then (F',G') is cofinal in (F,G).

Proof. Suppose $b \in F$. Then $b < a < G'$. Since a has minimal length among elements satisfying $F' < x < G'$ and b has smaller length than a, $F' < b$ is impossible, i.e. $(\exists c \in F')(c \geq b)$. This is precisely what we need. A similar argument applies to G and G'.

The same proof works if the representation in theorem 2.4 is used. At any rate, we now have what we need to build up the algebraic structure on the surreal numbers. It is hard to believe at this stage, but the relatively simple-minded system we have supports a rich algebraic structure.

3 THE BASIC OPERATIONS

A ADDITION

We define addition by induction on the natural sum of the lengths of the addends. Recall that the natural sum is obtained by expressing the ordinals in normal form in terms of sums of powers of ω and then using ordinary polynomial addition, in contrast to ordinary ordinal addition which has absorption. Thus the natural sum is a strictly increasing function of each addend.

The following notation will be convenient. If $a = F|G$ is the canonical representation of a, then a' is a typical element of F and a'' is a typical element of G. Hence $a' < a < a''$. We are now ready to give the definition.

Definition. $a + b = \{a'+b,\ a+b'\}|\{a''+b,\ a+b''\}$.

Several remarks are appropriate here. First, since the induction is on the natural sum of the lengths, we are permitted to use sums such as $a'+b$ in the definition. Secondly, no further definition is needed for the beginning. Since at the beginning we have only the empty set, we can use the trite remark that $\{f(x):x\varepsilon\phi\} = \phi$ regardless of f. For example, $\phi|\phi + \phi|\phi = \phi|\phi$. Thirdly, there is the a priori possibility that the sets F and G used in the definition of $a+b$ do not satisfy $F < G$. To make the definition formally precise, one can use the convention that $F|G = u$ for some special symbol u if $F \nless G$ and that $F|G = u$ if $u \varepsilon F \cup G$. In the sequel when a definition of an operation is given in the above form, we will show that F is always less than G so that the operation is really defined, i.e. u is never obtained as a value.

[1] is followed somewhat closely in building up the algebraic

operations. However, some differences are inevitable because of the different foundations. We have a specific system with a specific order. [1] deals with abstract elements and an order which is inductively defined by a method which corresponds to our theorem 2.5.

Note that since we use a specific representation of elements in the form $F|G$, the operations are automatically well defined. Nevertheless, in order to advance it is necessary to have the fact that the result is independent of the representation.

Let us illustrate the definition with several simple examples. Denote the empty sequence ( ) by 0 and the sequence (+) by 1. Now $(+) = \{0\}|\phi$. (It is easy to get confused. Note that G is the empty set and F is the unit set whose only element is the empty sequence. They are thus not the same.) Then $1+0 = \{0\}|\phi + \phi|\phi = \{0+\phi|\phi\}|\phi = \{0\}|\phi = 1$. Similarly $0+1 = 1$. Also $1+1 = \{0\}|\phi + \{0\}|\phi = \{0+1, 1+0\}|\phi = \{1\}|\phi = \{(+)\}|\phi = (++)$ which is natural to call "2".

It does look rather cumbersome to work directly with the definition, but so would ordinary arithmetic if we were forced to use $\{\phi\}$, $\{\phi,\{\phi\}\}$, instead of 1, 2, etc. and go back to inductive definitions.

Theorem 3.1. $a+b$ is always defined (i.e. never u) and furthermore $b > c \Rightarrow b > a + c$ and $b > c \Rightarrow b + a > c + a$.

Remark 1. Although the first part is what is most urgent, we need the second part to carry through the induction.

Remark 2. As a matter of style, one can prove commutativity first (which is trivial) and then simplify the statement of the above theorem and its proof. However, it seems preferable to prove that $a+b$ exists as a surreal number before proving any of its properties.

Proof. We use induction on the natural sum of the lengths. In other words, suppose theorem 3.1 is true for all pairs (a,b) of surreal numbers such that $\ell(a) + \ell(b)$ is less than α. We show that the statements remain valid if we include pairs whose natural sum is α.

Now $a + b = \{a'+b, a+b'\}|\{a''+b, a+b''\}$. First, we must show that $F < G$. Since $a' < a''$, it follows from the inductive hypothesis that $a' + b < a''+b$. Similarly $a + b' < a + b''$. Also $a' + b < a' + b'' < a + b''$ and $a + b' < a'' + b' < a'' + b$. Hence $a+b$ is defined.

By definition $a' + b < a + b < a'' + b$ and $a + b' < a + b < a + b''$. This proves the required inequality when either of b and c is an initial segment of the other.

Now suppose that neither b nor c is an initial segment of the other and such that $\ell(a) + \ell(b) \leq \alpha$ and $\ell(a) + \ell(c) \leq \alpha$. Let d be the common initial segment of b and c. Now assume $b > c$. Hence $b > d > c$. Hence $a + b > a + d > a + c$ and $b + a > d + a > c + a$.

It follows immediately that $a > b$ and $c > d \Rightarrow a + c > b + d$.

Theorem 3.2. Suppose $a = F|G$ and $b = H|K$; then $a + b = \{f+b, a+h\}|\{g+b, a+k\}$ where $f \in F$, $g \in G$, $h \in H$, $k \in K$. I.e. although the definition of $a + b$ is given in terms of the canonical representation of a and b, all representations give the same answer.

Remark. We shall call this "the uniformity theorem for addition," and say that the uniformity property holds for addition.

Proof. Let $a = F|G$, $b = H|K$. Suppose the canonical representations are $a = A'|A''$, $b = B'|B''$.

By the inverse cofinality theorem (theorem 2.9), F is cofinal in A' and similarly for the other sets involved. Consider $\{f+b, a+h\}|\{g+b, a+k\}$. It is now easy to check that the hypotheses of the cofinality theorem (theorem 2.6) are satisfied. The betweenness property of $a + b$ follows immediately from theorem 3.1, e.g. $f + b < a + b$. Also suppose $a'+b$ is one of the typical lower elements in the canonical representation of $a+b$ as in the definition. Since F is cofinal in A' $(\exists f \in F)(f \geq a')$. By theorem 3.1, $f + b \geq a' + b$. A similar argument applies to the other typical elements. Hence the cofinality condition is satisfied so by theorem 2.6 we do get $a+b$.

This technique will be used often to get uniformity theorems for other operations. Such results facilitate our work with these operations. In particular, they permit us to use the methods of [1] in dealing with composite operations, as we shall see, for example, in the proof of the associative law for addition.

Theorem 3.3. The surreal numbers form an Abelian group with respect to addition. The empty sequence is the identity, and the inverse is obtained by reversing all signs. (Note that one should be aware of potential set-theoretic problems since the system of surreal numbers is a proper class.)

Proof: 1) commutative law. This is trivial because of the symmetric nature of the definition.

2) associative law. We use induction on the natural sum of the lengths of the addends

$$(a+b) + c = \{(a+b)'+c,\ (a+b)+c'\}|\{(a+b)''+c,\ (a+b)+c''\}.$$

By theorem 3.2 we may use $a+b'$ and $b+a'$ instead of $(a+b)'$ and similarly for $(a+b)''$. i.e. it is convenient to use the representation in the definition of addition rather than the canonical representation. We thus obtain

$$(a+b) + c = \{(a'+b)+c,\ (a+b')+c,\ (a+b)+c'\}|\{(a''+b)+c,\ (a+b'')+c,\ (a+b)+c''\}.$$

A similar result is obtained for $a + (b+c)$. Associativity follows by induction.

3) The identity: Denote the empty sequence by 0. Then $0 = \phi|\phi$. We again use induction: $a + 0 = \{a'+0,\ a+0'\}|\{a''+0,\ a+0''\}$. There are no terms $0',0''$, so this simplifies to $\{a'+0\}|\{a''+0\}$ which is $\{a'\}|\{a''\}$ by the inductive hypothesis. We thus get $a + 0 = a$.

4) The inverse: We use induction. Let $-a$ be obtained from a by reversing all signs, and let $F|G$ be the canonical representation of a. Again let a' and a'' be typical elements of F and G respectively. Note that, in general, if b is an initial segment of c then $-b$ is an initial segment of $-c$ and $b < c \Rightarrow -b > -c$. Hence the canonical representation of $-a$ may be expressed as $-a''|-a'$. Therefore $a + (-a) = \{a'+(-a),\ a+(-a'')\}|\{a''+(-a),\ a+(-a')\}$. Since $a' < a < a''$, it is clear from the lexicographical order that $-a'' < -a < -a'$. Using induction and the fact that addition preserves order, we obtain

$a' + (-a) < a' + (-a') = 0$. $a + (-a'') < a'' + (-a'') = 0$.

$a'' + (-a) > a'' + (-a'') = 0$, $a + (-a') > a' + (-a') = 0$. Hence in the representation of $a + (-a)$, as $H|K$, $H < 0 < K$. Since 0 vacuously satisfies any minimality property, $a + (-a) = H|K = 0$.

Thus we now know that the surreal numbers form an ordered Abelian group. The identity and inverses are obtained in a way which is heuristically natural.

B MULTIPLICATION

The definition of multiplication is more complicated than that of addition; in fact, it took some time before the standard definition for multiplication was discovered.

Definition. $ab = \{a'b + ab' - a'b', a''b + ab'' - a''b''\}|$ $\{a'b + ab'' - a'b'', a''b + ab' - a''b'\}$.

As partial motivation note that if a,b,a',b' are ordinary real numbers such that $a' < a$, $b' < b$, then $(b-b')(a-a') > 0$, i.e. $a'b + ab' - a'b' < ab$. Similar computations apply to get the appropriate inequalities if either a' is replaced by a'' or b' replaced by b''. Thus the inequalities are consistent with what is desired.

Theorem 3.4. ab is always defined. Furthermore $a > b$ and $c > d \Rightarrow ac-bc > ad-bd$.

Proof. We use induction on the natural sum of the lengths of the factors. We shall refer to the inequalities $ac - bc > ad - bd$ as $P(a,b,c,d)$ and to the expression $a^o b + ab^o - a^o b^o$ where a^o, b^o are proper initial segments of a and b respectively as $f(a^o,b^o)$. Note that a^o is of the form a' or a''. In the former case we call a^o a lower element and in the latter case an upper element.

It follows immediately from the definition that the relation P is transitive on the last two variables. Since, at this point, we can freely use the properties of addition in ordinary algebra $P(a,b,c,d)$ is equivalent to $ac - ad > bc - bd$. This makes it clear that P is transitive on the first two variables.

Now let b_1^0 and b_2^o be initial segments of b and consider $f(a^o,b_2^o) - f(a^o,b_1^0)$. This is $(a^o b+ab_2^o-a^o b_2^o) - (a^o b+ab_1^0-a^o b_1^0) = (ab_2^o-a^o b_2^o) - (ab_1^0-a^o b_1^0)$. If $a > a^o$ and $b_2^o > b_1^0$, the inductive hypothesis says that the above expression is positive, i.e. $P(a,a^o,b_2^o,b_1^0)$, so $f(a^o,b^o)$ is an increasing function of b^o if $a^o < a$. If $a^o > a$ and $b_1^0 > b_2^o$, the above expression may be written $(a^o b_1{}^o - ab_1{}^o) - (a^o b_2{}^o - ab_2{}^o)$ which, again, is positive. Hence if $a^o > a$, $f(a^o,b^o)$ is a decreasing function of b^o. Similarly, $f(a^o,b^o)$ is an increasing function of a^o if $b^o < b$ and decreasing if $b^o > b$.

To show that ab is defined, we must check inequalities such as $f(a_1',b') < f(a_2',b'')$. This follows easily from the above. If $a_1' = a_2'$, this is immediate since $b' < b''$ and $a_1' < a$. More generally, let a' be max (a_1',a_2'); then $f(a_1',b') \leq f(a',b') < f(a',b'') \leq f(a_2',b'')$. We now consider $f(a_1'',b'')$ and $f(a_2'',b')$ and let $a'' = \min(a_1'',a_2'')$. Then $f(a_1'',b'') \leq f(a'',b'') < f(a'',b') \leq f(a_2'',b')$. Since the remaining cases are similar to the ones we already checked, this shows that ab is defined.

To prove the second statement of the theorem, assume first that in each of the pairs $\{a,b\},\{c,d\}$ one element is an initial segment of the other. By definition $f(a',b') < ab$, $f(a'',b'') < ab$, $f(a',b'') > ab$ and $f(a'',b') > ab$. Now $ab - f(a',b') = ab - (a'b+ab'-a'b')$ $= (ab-a'b) - (ab'-a'b')$. Thus the first inequality says $P(a,a',b,b')$. Similarly, the other inequalities give $P(a'',a,b'',b)$, $P(a,a',b'',b)$, $P(a'',a,b,b')$. This proves the statement in this special case.

Next, remove the above restriction on $\{a,b\}$, but still assume that one of c,d is an initial segment of the other. Let e be the common initial segment of a and b. Then $a > e > b$. By the above $P(a,e,c,d)$ and $P(e,b,c,d)$. Hence by transitivity we obtain $P(a,b,c,d)$.

Finally, suppose neither c nor d is an initial segment of the other and let f be their common initial segment. By the above, we have $P(a,b,c,f)$ and $P(a,b,f,d)$, so we finally obtain $P(a,b,c,d)$ by transitivity. This takes care of all cases.

Theorem 3.5 (The uniformity theorem for multiplication). The uniformity property holds for multiplication.

Proof. This is similar to the proof of theorem 3.2 and, in fact, all theorems of this type have a similar proof once we have basic inequalities of a suitable kind.

Now that we have theorem 3.4 and, in particular, the fact that the inequality stated there is valid in general, the same computation as in the beginning of the proof of the theorem gives us the behaviour of $f(c^o,d^o)$ in general for $c^o \neq a$ and $d^o \neq b$. (We no longer require that c^o and d^o be initial segments of a and b respectively.)

Suppose $a = F|G$, $b = F'|G'$, $c^o \in F \cup G$, $d^o \in F' \cup G'$. Then $f(c^o,d^o)$ is an increasing function of d^o if $c^o < a$ and a decreasing function of d^o for $c^o > a$ and similarly for fixed d^o.

We now check the hypotheses of the cofinality theorem. The betweenness property of ab follows from the same computation as in the latter part of the proof of theorem 3.4. For example, since $ab - f(c',d') = (ab-c'b) - (ad'-c'd')$, $P(a,c',b,d')$ says that $ab > f(c',d')$. The other parts of the betweenness property follow the same way. Note that we are now going in the reverse direction to the one we went earlier, i.e. we have P and we obtain the betweenness property.

To check cofinality let e.g. $f(a',b')$ be a lower element using the canonical representation of ab. By the inverse cofinality theorem $(\exists c \in F)(c \geq a')$ and $(\exists d \in F')(d \geq b')$. Then $f(c,d) \geq f(c,b') \geq f(a',b')$. A similar argument applies to the other cases. Actually all the cases may be elegantly unified by noting that $f(c^o,x)$ maintains the side of x if $c^o < a$ and reverses it if $c^o > a$. Thus $f(c^o,x)$ maintains the side of x if and only if it is an increasing function of x. Hence in all cases $f(c^o,x)$ is closer to ab if x is closer to b. Similarly for $f(x,d^o)$. This is just what is needed to obtain cofinality.

Theorem 3.6. The surreal numbers form an ordered commutative ring with identity with respect to the above definitions of addition and multiplication. The multiplicative identity 1 is the sequence $(+) = \{0\}|\phi$.

Proof. Commutative law. Because of symmetry this is just as trivial here as it was for addition.

Distributive law. We use induction on $\ell(a) + \ell(b) + \ell(c)$.
$(a+b)c = \{(a+b)'c + (a+b)c' - (a+b)'c', \cdots\}|\cdots$. By theorem 3.5 we may use $a+b'$ and $a'+b$ instead of $(a+b)'$. For unification purposes we consider a typical term $(a+b)^oc + (a+b)c^o - (a+b)^oc^o$ in the representation of $a(b+c)$ where an element is lower if and only if an even

number of noughts correspond to double primes. For $(a+b)^o$ we use $\{a^o+b, a+b^o\}$ by the above. Thus, typical terms become $(a^o+b)c + (a+b)c^o - (a^o+b)c^o$ and $(a+b^o)c + (a+b)c^o - (a+b^o)c^o$. By induction the first term becomes $a^oc + bc + ac^o + bc^o - a^oc^o - bc^o$ $= a^oc + ac^o - a^oc^o + bc$. A similar result is obtained if $a + b^o$ is used instead of $a^o + b$.

A typical term in the representation of $ac+bc$ is $(ac)^o + bc$ which is $(a^oc+ac^o-a^oc^o) + bc$. Note that this is justified by theorem 3.2. Since a similar result applies if we take $ac + (bc)^o$ and since the parity rule as to which element is upper or lower is the same as before, we see that $(a+b)c = ac+bc$.

We are now permitted to write $ab - (a^ob+ab^o-a^ob^o)$ as $(a-a^o)(b-b^o)$.

Associative law. We use induction on $\ell(a) + \ell(b) + \ell(c)$. A typical term in the representation of $(ab)c$ is $(ab)^oc + (ab)c^o - (ab)^oc^o$ which by theorem 3.5 may be written $(a^ob+ab^o-a^ob^o)c + (ab)c^o - (a^ob+ab^o-a^ob^o)c^o$. Again, an element is lower if and only if an even number of noughts correspond to double primes. By the distributive law the above expression is $(a^ob)c + (ab^o)c - (a^ob^o)c + (ab)c^o - (a^ob)c^o - (ab^o)c^o + (a^ob^o)c^o$. Of crucial importance is the following kind of symmetry in the expression: the terms are all obtained from $(ab)c$ by putting a superscript on at least one of the factors, and the term has a minus if and only if there is an even number of superscripts.

Exactly the same thing happens with the expansion of $a(bc)$ except for the bracketing. I.e. the parity rules as to which term is upper or lower, and which addends in a term have a minus is the same as before. In fact, we obtain for a typical term

$$a(bc) = a^o(bc) + a(b^oc+bc^o-b^oc^o) - a^o(b^oc+bc^o-b^oc^o)$$
$$= a^o(bc) + a(b^oc) + a(bc^o) - a(b^oc^o) - a^o(b^oc) - a^o(bc^o) + a^o(b^oc^o).$$

The result now follows by induction.

The identity. First note that $a \cdot 0 = 0$. This follows from the distributive law. It also follows immediately from the definition. Since $0 = \phi|\phi$, and all terms used in representing a product must contain lower or upper representatives of each factor, $a \cdot 0 = \phi|\phi = 0$. (Note that this was not the case for addition.)

We again use induction to compute $a \cdot 1$. Since $1 = \{0\}|\phi$ the expression for $a \cdot 1$ reduces to $a \cdot 1 = \{a' \cdot 1 + a \cdot 0 - a' \cdot 0\}|\{a'' \cdot 1 + a \cdot 0 - a'' \cdot 0\}$ which is $\{a' \cdot 1\}|\{a'' \cdot 1\}|$. By induction this is $\{a'|a''\}$ which is a.

Compatibility of ordering. Suppose $a > 0$ and $b > 0$. Then by theorem 3.4 we have $P(a,0,b,0)$, i.e. $a \cdot b - 0 \cdot b > a \cdot 0 - 0 \cdot 0$. Hence $ab > 0$.

Thus we now know that the surreal numbers form an integral domain. We saw that multiplication behaves somewhat more subtly than addition. We shall see in the next section that division is handled much more subtly. At any rate, it is remarkable that all this is possible.

It is possible to get a nice form for the representation of an n-fold sum and product by inductive use of the uniformity theorems. It is trivial that $a_1 + a_2 + \cdots + a_n$ may be represented as
$\{a_1' + a_2 \cdots + a_n, a_1 + a_2' \cdots + a_n, \cdots a_1 + a_2 \cdots + a_n'\}|$
$\{a_1'' + a_2 \cdots + a_n, a_1 + a_2'' \cdots + a_n, a_1 + a_2 \cdots + a_n''\}$.

We claim that $a_1 a_2 \cdots a_n$ may be represented by terms $a_1 a_2 \ldots a_n - (a_1 - a_1^0)(a_2 - a_2^o) \cdots (a_n - a_n^o)$ where an element is lower if and only if an even number of noughts correspond to double primes.

The identity $ab - (a^o b + a b^o - a^o b^o) = (a - a^o)(b - b^o)$ may be written in the succinct form $ab - (ab)^o = (a - a^o)(b - b^o)$. Theorem 3.5 allows us to use this representation for ab if we multiply by other factors. Thus it is clear by induction that
$(a_1 a_2 \ldots a_n) - (a_1 a_2 \ldots a_n)^o = (a_1 - a_1^0)(a_2 - a_2^o) \ldots (a_n - a_n^o)$. The parity rule is easily checked. In fact, it is essentially the same as the one in ordinary algebra for multiplying pluses and minuses.

It is possible to use the above computation to prove the associative law. Of course, one would have to be more cautious with the bracketing before one has that law.

C DIVISION

We will define a reciprocal for all $a > 0$. As usual, induction will be used, but the definition is more involved than the ones for the earlier operations. Let $a = A'|A''$ be the canonical representation. One naive attempt would be as follows. We try $x = \{0, \frac{1}{a''}\}|\{\frac{1}{a'}\}$ where $a'' \in G$ and $a' \in F - \{0\}$. (Note that $0 \in A'$ since $a > 0$.) Unfortunately this does not work. Although x has some of the properties expected of a candidate for $\frac{1}{a}$, $xa \neq 1$ in general. It turns out that more elements are needed to get a representation for the reciprocal. Roughly speaking, the idea is to insert as many elements into the representation of x as is needed to force the crucial inequalities, i.e. in the standard representation of ax as a product we want the lower elements to be less than one and the upper elements to be greater than one.

What is needed is more complicated. First we define objects $\langle a_1, a_2, \cdots, a_n \rangle$ for every finite sequence where $a_i \in A'$ $A'' - \{0\}$. For arbitrary b we define $b \circ a_i$ as the unique solution of $(a-a_i)b + a_i x = 1$. This exists by the inductive hypothesis which guarantees that a_i as an initial segment of a has an inverse. Uniqueness is automatic. Now let $\langle\ \rangle = 0$ and $\langle a_1, a_2, \cdots, a_{n+1} \rangle = \langle a_1, a_2, \cdots, a_n \rangle \circ a_{n+1}$ For example $\langle a_1 \rangle = 0 \circ a_1 = a_1^{-1}$. We now claim that $a^{-1} = F|G$ where $F = (\langle a_1, \cdots, a_n \rangle$: the number of $a_i \in A'$ is even) and $G = (\langle a_1, \cdots, a_n \rangle$: the number of $a_i \in A'$ is odd).

Theorem 3.7. The surreal numbers form a field.

Proof. We first show that $x \in F \Rightarrow ax < 1$ and $x \in G \Rightarrow ax > 1$. This will show that $F < G$. Since $\langle\ \rangle \in F$, $\langle\ \rangle = 0$, and $a \cdot 0 = 0 < 1$, the result is valid for $\langle\ \rangle$. We now use induction on the length of the finite sequence. In other words it is enough to show that if b has this property so does $x = b \circ a_1$. Now by definition $(a-a_1)b + a_1 x = 1$. Clearly $(a-a_1)b + a_1 b = ab$. Since $a_1 > 0$ it follows that $x > b$ iff $1 > ab$. Also it follows from the above identity that $ax = 1 + (a-a_i)(x-b)$.

Now for fixed a_1, the map $b \to b \circ a_1$ preserves being in F or G iff a_1 is upper. For example, $b \in F$ and $a_1 \in A' \to b \circ a_1 \in G$. In that case $ab < 1$ by the inductive hypothesis, thus $x > b$. Since $a > a_1$ it follows that $ax = 1 + (a-a_1)(x-b) > 1$. Hence $x \in G$ and $ax > 1$ as desired. The other three cases can be unified by noting that any change in b or a_1 from lower to upper leads to a change in $b \circ a_1$ and hence reverses the desired inequality. At the same time, any change in a_1 or b reverses the sign of $a-a_1$ and $x-b$ respectively so it also reverses the inequality we actually have. This proves what we desired, so that now we know that $F < G$ and therefore $F|G$ has meaning. Let $F|G = c$.

Finally we compute ac. A typical element used in the definition of the product has the form $a_1c + ac_1 - a_1c_1$ where $a_1 \in A' \cup A''$ and $c_1 \in F \cup G$. First, $0 \in A'$ and $0 \in F$. Thus we get a lower element in the representation of ac by choosing $a_1 = c = 0$. Hence $0 \cdot c + a \cdot 0 - 0 \cdot 0 = 0$ is a lower element.

Suppose $a_1 = 0$. Then the elements in the definition of the product reduce to ac_1 and ac_1 is an upper element for ac iff $c_1 \in G$.

However, we know that $c_1 \in G \rightarrow ac_1 > 1$ and $c_1 \in F \rightarrow ac_1 < 1$. Hence if ac_1 is an upper element $ac_1 > 1$ and if ac_1 is a lower element $ac_1 < 1$.

Now suppose $a_1 \neq 0$. Then $c_1 \circ a_1$ is defined, is contained in $F \cup G$, and satisfies the equation $(a-a_1)c_1 + a_1x = 1$. Now $a_1c + ac_1 - a_1c_1$ is a lower element for ac iff a_1 and c_1 are on the same side of a and c respectively iff $c_1 \circ a_1 \in G$ iff $c_1 \circ a_1 > c$. (This follows from the earlier statement regarding the map $b \rightarrow b \circ a_1$.) Since $c_1 \circ a_1$ satisfies the equation $(a-a_1)c_1 + a_1x = 1$ and $a_1 > 0$ the inequality $c_1 \circ a_1 > c$ is equivalent to $(a-a_1)c_1 + a_1c < 1$. The left-hand side of this inequality is nothing but $a_1c + ac_1 - a_1c_1$. Hence the lower elements for ac are less than 1 and the upper elements are greater than 1. (Note that since $c_1 \circ a_1 \in F \cup G$, it follows that $c_1 \circ a_1 \neq c$ so that the negation of ">" may be taken to be "<" in the proof.)

We have shown that in the expression for ac, 1 satisfies the betweenness property but 0 being in the lower part does not. Hence $1 = (+)$ is the number of minimal length satisfying the betweenness property. Therefore $ac = 1$.

Thus finally we know that we have a field. Although we don't need the information it is of passing interest to note how $b \circ a_1$ varies as a function of b and a_1. First, by solving the defining equation for $b \circ a_1$ we get $b \circ a_1 = \dfrac{1-(a-a_1)b}{a_1} = b + \dfrac{1-ab}{a_1}$. The first expression implies that $b \circ a_1$ is an increasing function of b iff $a < a_1$ and the second expression that $b \circ a_1$ is an increasing function of a_1 iff $ab > 1$. Hence $b \circ a_1$ is an increasing function of one of the variables iff the other variable is upper. At the same time the function preserves sides iff the fixed variable is upper. All this can be unified by saying that $b \circ a_1$ gets closer to c if b gets closer to c and if a_1 gets closer to a. As we already saw in dealing with addition and multiplication, this is essentially what is needed to prove a uniformity theorem. Since the details are routine and since we don't need it, this will not be pursued.

Finally it is recommended to any reader who is confused by the unified arguments to think first in terms of individual cases, e.g. in the above assume b and a_1 are upper and regard $b \circ a_1$ as a function of b for fixed a_1.

D. SQUARE ROOT

Frankly, it is not of extreme importance to obtain the existence of square roots at this time, since that can be obtained from the theory of infinite series which will be developed later. However, in view of the elegance of the theory, it is worth seeing how square roots can be obtained directly without further machinery. In the bottom of page 22 in [1] credit for this is given to Clive Bach.

We assume $a > 0$ and use induction. I.e. we assume that all initial segments of a (they are necessarily non-negative) have square roots. Let $a = A'|A''$ be the canonical representation. Then all elements in $A' \cup A''$ have square roots. Let H be the free groupoid, with product denoted by $\circ$, generated by the elements of $A' \cup A''$. We shall define inductively a partial map from H into the surreal numbers.

If $b \in A' \cup A''$ then $f(b) = \sqrt{b}$.

If $b, c \in H$, $f(b)$ and $f(c)$ are defined and are not both 0, then $f(b \circ c) = \dfrac{a+f(b)f(c)}{f(b)+f(c)}$.

(In analogy with this case it was possible to use the concept of free semigroup to deal with division, but we preferred to be more concrete. Here we are stuck with this formalism since we are dealing with non-associative juxtaposition.) By induction $(\forall x)[f(x) \geq 0]$. Furthermore $f(x) > 0$ unless $x = 0$.

F and G are now defined inductively as follows.

If $b \in A'$ then $f(b) \in F$. If $b \in A''$ then $f(b) \in G$. $f(b \circ c) \in G$ if $f(b)$ and $f(c)$ are both in F or both in G. If one of $f(b)$ and $f(c)$ is in F and the other in G, then $f(b \circ c) \in F$. Since we are not assuming that f is one-one, a priori, it may seem possible that F and G have elements in common. However, we shall prove that $F < G$ which in particular guarantees that F and G are disjoint.

Theorem 3.8. Every positive element has a square root.

Proof. We will show that $F|G = \sqrt{a}$. First we show that $x \in F \Rightarrow x^2 < a$ and $x \in G \Rightarrow x^2 > a$. This is clearly true for $x \in A' \cup A''$. In order to carry through an induction it is necessary to study the behaviour of

$x \circ y = \dfrac{a+xy}{x+y}$ as a function of x and y. First

$x_1 \circ y - x_2 \circ y = \dfrac{(y^2-a)(x_1-x_2)}{(x_1+y)(x_2+y)}$. Hence, for fixed y, $x \circ y$ is an increasing function of x iff $y^2 > a$ iff $y \in G$ using induction. Also $y \in G$ iff the map $x \to x \circ y$ preserves presence in F or G according to the above definition. Thus we have a similar desired situation to one we have previously, i.e. preserving sides is equivalent to being an increasing function if one variable is fixed.

Now assume $x_1 \circ x_2 \in G$. First, if $x_1, x_2 \in F$, let $x = \max(x_1, x_2)$. Then, by the above $x_1 \circ x_2 \geq x \circ x$. Similarly, if $x_1, x_2 \in G$ we take $x = \min(x_1, x_2)$ and obtain $x_1 \circ x_2 \geq x \circ x$. Now $x \circ x = \dfrac{a+x^2}{2x}$.

By the inductive hypothesis, if $x \in F$, then $x^2 < a$ and if $x \in G$, then $x^2 > a$. In either case, $x^2 \neq a$. Hence $(a-x^2)^2 > 0$. Thus $a^2 + 2ax^2 + x^4 > 4ax^2$. Therefore $\left(\dfrac{a+x^2}{2x}\right)^2 = \dfrac{a^2+2ax^2+x^4}{4x^2} > a$. A fortiori $(x_1 \circ x_2)^2 > a$.

Now assume $x \circ y \in F$. Without loss of generality (because $x \circ y$ is symmetric in x and y) suppose $x \in F$ and $y \in G$. Then $x^2 < a$ and $y^2 > a$.

Assume first that $xy = a$. Then $x \circ y = \dfrac{a+xy}{x+y} = \dfrac{2xy}{x+y}$ and $(x \circ y)^2 = \dfrac{4xya}{(x+y)^2}$. Clearly $x \neq y$ since $x^2 < a < y^2$; hence $\dfrac{4xy}{(x+y)^2} < 1$. Therefore $(x \circ y)^2 < a$ as desired.

If $xy \neq a$, then either $x < \frac{a}{y}$ or $x > \frac{a}{y}$. If $x < \frac{a}{y}$, we apply the above to $\frac{a}{y}$ and y to obtain $(\frac{a}{y} \circ y)^2 < a$. The above applies since membership in F or G is not required in the argument. All we want is that $(\frac{a}{y})^2 < a$. (Even the latter is not really needed since we can get by even if all we have is $(\frac{a}{y} \circ y)^2 \leq a$.) Since $y^2 > a$, $x \circ y$ is an increasing function of x and since $x < \frac{a}{y}$, we have

$(x \circ y)^2 < (\frac{a}{y} \circ y)^2 < a$.

If $x > \frac{a}{y}$, then $y > \frac{a}{x}$ (x is necessarily different from 0). Then we apply the earlier argument to x and $\frac{a}{x}$ and obtain $(x \circ \frac{a}{x})^2 < a$. Since $x^2 < a$, $x \circ y$ is a decreasing function of y. Hence

$(x \circ y)^2 < (x \circ \frac{a}{x})^2 < a$.

This finally shows that $x \in F \Rightarrow x^2 < a$ and $x \in G \Rightarrow x^2 > a$. Since $x \geq 0$ this shows that $F < G$. Hence $F|G$ has meaning. Let $F|G = c$.

We now compute c^2. Then a typical term in the representation of c^2 is $c_1c + c_2c - c_1c_2$. This is lower iff c_1 and c_2 are on the same side of c iff $c_1 \circ c_2$ is an upper element, i.e.

$c < c_1 \circ c_2 = \dfrac{a+c_1c_2}{c_1+c_2}$ iff $c(c_1+c_2) < a + c_1c_2$ iff $c_1c + c_2c - c_1c_2 < a$.

The argument breaks down if $c_1 = c_2 = 0$ since $c_1 \circ c_2$ is undefined. But this case leads to a lower element which is 0 which is less than a. So lower elements are less than a and upper elements greater than a, i.e. a satisfies the betweenness property for c^2.

Now $0, \sqrt{a'} \in F$. Hence one of the lower elements in the representation of c^2 is $c\sqrt{a'} + c(0) - (\sqrt{a'})(0) = c\sqrt{a'} \geq \sqrt{a'}\sqrt{a'} = a'$. Now $\sqrt{a''} \in G$. Hence one of the upper elements is

$c\sqrt{a''} + c(0) - (\sqrt{a''})(0) = c\sqrt{a''} \leq \sqrt{a''}\sqrt{a''} = a''$.

By the cofinality theorem $c^2 = a$.

Note that as in the case of division, what we did was to insert just enough terms into F and G in order to force the betweenness condition. Again, just as in the case of division, we have what is needed to prove a uniformity theorem.

4 REAL NUMBERS AND ORDINALS

A INTEGERS

The main task of this chapter is to show that the surreal numbers contain both the real numbers and the ordinals. (The distinction as to whether the surreal numbers contain the real numbers or a set isomorphic to the real numbers is very much like the distinction as to whether the Iliad was written by Homer or by someone else of the same name.) Along the way we shall see the explicit representation of ordinary numbers as sequences of pluses and minuses. So far we know that the additive identity 0 is the empty sequence and that the multiplicative identity 1 is the sequence (+). Since the surreal numbers form an ordered field, the expression $(\overbrace{1+1+1\cdots}^{n\text{ times}})$ may be identified with the the positive integer n. We now have the following result which is consistent with one's heuristic expectations.

<u>Theorem 4.1</u>. The positive integer n is $(\overbrace{+++\cdots}^{n\text{ times}})$.

<u>Proof</u>. We use complete induction, i.e. suppose the theorem is true for all integers $m \leq n$. Then $\overbrace{1+1+1\cdots}^{n+1\text{ times}} = (\overbrace{1+1+1\cdots}^{n\text{ times}}) + 1$ $= (\overbrace{+++\cdots}^{n\text{ times}}) + (+)$ by the inductive hypothesis. By applying cofinality to the canonical representation we know that $(\overbrace{+++\cdots}^{n\text{ times}})$ may be expressed as $(\overbrace{+++\cdots}^{n-1\text{ times}})|\phi = \{n-1\}|\phi$ by the inductive hypothesis. (+) is clearly $\{0\}|\phi$. Hence by definition $n+1 = \{(n-1)+1, n+0\}|\phi = \{n\}|\phi = (\overbrace{+\cdots}^{n\text{ times}})|\phi$ which is $(\overbrace{+++\cdots}^{n+1\text{ times}})$ again by applying cofinality to the canonical representation of $(\overbrace{+++\cdots}^{n+1\text{ times}})$.

Note that the symbol + has been used in two different senses, once for addition, and once as one of the symbols used in the ordinal sequences which we consider. This happens also in ordinary algebra where, for example, in +(a+b) the two pluses have different meanings. However, the meanings are related in such a way that it is convenient in practice to use the same symbol for each. In our case the use of the symbols + and - is consistent with the intuitive feeling of order, i.e. plus is above zero is above minus. In any case, the meaning should be clear from the context.

Corollary. The negative integer $-n$ is $(\overbrace{---\cdots}^{n \text{ times}})$.

Proof. This is an immediate consequence of the theorem and the formula for the additive inverse obtained previously.

B DYADIC FRACTIONS

Since the class of surreal numbers contain the rational numbers, it seems natural to consider them next and even to conjecture that the rational numbers correspond to finite sequences of pluses and minuses. Since $0 = (\ ) < (+-) < (+) = 1$, it is natural to conjecture that $(+-) = \frac{1}{2}$. A heuristic guess for $(+--)$ would be a toss-up between $\frac{1}{3}$ and $\frac{1}{4}$. Actually $(+-) = \frac{1}{2}$ and $(+--) = \frac{1}{4}$.

It turns out that the finite sequences correspond to the dyadic fractions, i.e. rationals of the form $\frac{a}{2^n}$. Although they form a proper subset of the rationals, they are dense in the reals. Thus they can be used just as well as the rationals as building blocks later in developing the reals.

Lemma 4.2. If $\{2a\}|\{2b\} = a+b$ then $\{a\}|\{b\} = \frac{a+b}{2}$.

Proof. Let $\{a\}|\{b\} = c$. Then $2c = c+c = \{a+c\}|\{b+c\}$ by definition of addition. We show that the latter is $a+b$ by cofinality. First, $a < c < b$. Hence $a + c < a + b < b + c$ which is the betweenness property. Now $\{2a\}|\{2b\} = a+b$. Also it follows from $a < c < b$ that $2a < a + c$ and $b + c < 2b$. Thus we have the cofinality property. So $2c = a+b$; therefore $\{a\}|\{b\} = c = \frac{a+b}{2}$.

The above is the key lemma for dealing with dyadic fractions.

For example $1 = \{0\}|\phi = \{0\}|\{2\}$ by cofinality. Hence the hypothesis of lemma 4.2 is satisfied if $a = 0$ and $b = 1$. Hence $\{0\}|\{1\} = \frac{1}{2}$. So $\frac{1}{2} = \{(\ )\}|\{(+)\} = (+-)$. This says that the hypothesis of the lemma is valid for $a = 0$ and $b = \frac{1}{2}$. Hence $\{0\}|\{\frac{1}{2}\} = \frac{1}{4}$. So $\frac{1}{4} = \{(\ )|\{(+-)\} = \{(+--)\}$. This sets up an induction, but only numbers of the form $\frac{1}{2^n}$ will be reached. However, we also have, for example, $3 = \{0,1,2\}|\phi = \{2\}|\{4\}$ by cofinality. Again, by the lemma we obtain $\frac{3}{2} = \{1\}|\{2\} = \{(+)\}|\{(++)\} = (++-)$.

We now show that this process enables us to determine the rational number which corresponds to an arbitrary surreal number of finite length.

Theorem 4.2. Surreal numbers of finite length correspond to dyadic fractions.

Specifically, let d be a surreal number of length $m+n$ which satisfies

$$i,j < m \Rightarrow d(i) = d(j)$$

$$d(m) \neq d(0).$$

Define $b(i)$ as follows.

$b(i) = 1$ if $i < m$ and $d(i) = +$.

$b(i) = -1$ if $i < m$ and $d(i) = -$.

$b(i) = \frac{1}{2^{i-m+1}}$ if $i \geq m$ and $d(i) = +$.

$b(i) = -\frac{1}{2^{i-m+1}}$ if $i \geq m$ and $d(i) = -$.

Then $d = \sum_{i=0}^{m+n-1} b(i)$.

Remark. The above says informally that a plus is counted as a 1 and a minus as -1 until a change in sign occurs at which point the sequence of pluses and minuses is treated like a binary decimal (with 1 and -1 rather than with 1 and 0.) For example,

$+++-+-$ is $3 - \frac{1}{2} + \frac{1}{4} - \frac{1}{8} = 2\frac{5}{8}$.

Proof. Let $d(0) = +$. A similar argument applies if $d(0) = -$. (As an alternative one can take the result for $d(0) = +$ and take the negative of both sides.)

The case $n = 0$, which is the case where there is no change in sign, is essentially the statement of theorem 4.1.

We do the case $n = 1$ individually since this case is special. Here we have $d(m) = -$. The sequence consists of m pluses followed by a minus. To avoid confusion recall that the ordinals begin with 0 and the length is the least ordinal for which d is not defined. (This may seem unnatural in the finite case, but is required if one wants a general definition.) In any case, the two "unnatural" conventions cancel so that the length is really the number of terms in the sequence!

Of course $m \geq 1$ $[d(m) \neq d(0)]$. Now $2m-1 = \{0,1,2,\cdots 2m-2\}|\phi$. This is the canonical representation by theorem 4.1. By cofinality we obtain $2m-1 = \{2m-2\}|\{2m\}$.

We can now apply lemma 4.2 with $a = m-1$ and $b = m$ to obtain $\{m\}|\{m-1\} = m - \frac{1}{2}$. It is easy to see directly from the definition that $\{m\}|\{m-1\} = d$. $m-1$ consists of $(m-1)$ pluses and m of m pluses. Any surreal number between $m-1$ and m must by the lexicographical order begin with m pluses followed by a minus, i.e. have d as an initial segment. Hence $d = m - \frac{1}{2}$, which is exactly what the theorem says.

We now use induction on n. Assume that the theorem is true for all $n \leq r$ and let $n = r+1$.

We first note that an immediate application of induction to lemma 4.2 shows that $\{2a\}|\{2b\} = a+b \rightarrow \{\frac{a}{2^s}\}|\{\frac{b}{2^s}\} = \frac{a+b}{2^{s+1}}$ for all positive integers s. We already noted that the hypothesis is valid for consecutive integers c and $c+1$. Hence $\{\frac{c}{2^s}\}|\{\frac{c+1}{2^s}\} = \frac{c}{2^s} + \frac{1}{2^{s+1}}$.

Let d' be the initial segment of d of length $m+r$. Then $d = d'+$ or $d'-$ since a has length $m+n = m+r+1$. Assume $d = d'+$. (A similar argument applies if $d = d'-$.) Since the case $n = 1$ has been done separately, we can assume that $r \geq 1$, i.e. that d' begins with m pluses followed by a minus.

Now let $d = F|G$ be the canonical representation. F and G are finite, so by cofinality a has the form $\{x\}|\{y\}$ where x is the largest element in F and y the smallest element in G. Since $d = d'+$, clearly $x = d'$. We cannot be as explicit with y, since in this general situation we have very little information about the minuses in G. We know that there is a minus after m pluses; hence $y \leq m$. Also, we can apply the inductive hypothesis to x and y. Hence $d' = x = \frac{c}{2^r}$ for some integer c. If we can show that $y = \frac{c+1}{2^r}$ then we can apply the above formula to obtain $d = \{x\}|\{y\} = \frac{c}{2^s} + \frac{1}{2^{s+1}}$ which is exactly what we need to prove the theorem.

Our apparent lack of control over y will be overcome by the box principle. Let H be the set of all surreal numbers of length not greater than m+r that begin with m pluses followed by a minus. The cardinality of H is $1 + 2 + 2^2 \cdots 2^{r-1}$. By the inductive hypothesis every element of H is of the form $\frac{k}{2^r}$ for some integer k and by the lexicographical order is strictly between m-1 and m. Since there are precisely 2^r-1 such numbers, by the box principle <u>every</u> number of the form $\frac{k}{2^r}$ strictly between m-1 and m is in H. In particular, $\frac{c+1}{2^r} \in H$ unless $\frac{c+1}{2^r} = m$. In either case $\ell(\frac{c+1}{2^r}) \leq m+r$. Now $\frac{c+1}{2^r} > \frac{c}{2^r} = d'$. Since $d = d'+$ it follows from the lexicographical order that $\frac{c+1}{2^r} > d$. Now $G \subset H \cup \{m\}$. Hence every element of G has the form $\frac{k}{2^r}$. Since $\frac{c}{2^r} = d' < d$ and $d < G$, $\frac{c+1}{2^r}$ is a lower bound to G. In fact, $\frac{c+1}{2^r}$ is actually an initial segment of d. Otherwise, by considering the common initial segment of d and $\frac{c+1}{2^r}$ we obtain an element of G below $\frac{c+1}{2^r}$ which is a contradiction. Since $\frac{c+1}{2^r} > d$, it follows that $\frac{c+1}{2^r} \in G$ and hence the least element of G, i.e. $y = \frac{c+1}{2^r}$. As we said earlier, this is what we need to complete the proof.

During the proof we showed that all dyadic fractions are obtained this way. Also it is easy to see how to express a dyadic fraction constructively as a sequence. Heuristically speaking, we always go in the right direction to close in on the fraction. For example, consider $2\frac{3}{8}$. Since $2 < 2\frac{3}{8} < 3$, we begin with +++-. This is $2\frac{1}{2}$. Since $2\frac{3}{8} < 2\frac{1}{2}$ we want a minus next. Now +++-- is $2\frac{1}{4}$ so we need a +. Finally, +++--+ hits what we desire on the nose. More formally, one can set up an elementary induction. We assume that all fractions of the form $\frac{a}{2^n}$ with a odd correspond to sequences of length m+n. Consider

$\frac{2b+1}{2^{n+1}}$. Precisely one of b and $b+1$ is odd.

$\frac{2b+1}{2^{n+1}} = \frac{b}{2^n} + \frac{1}{2^{n+1}} = \frac{b+1}{2^n} - \frac{1}{2^{n+1}}$. Thus depending on which is odd we take the corresponding sequence of length $m+n$ and juxtapose a plus or minus, e.g. if b is odd we juxtapose a plus to the sequence of b. Note the lack of choice. Since $b+1$ is even in that case, $\frac{b+1}{2^n}$ is a sequence of length less than $m+n$. Tacking on a minus will thus not give $\frac{2b+1}{2^{n+1}}$. This is, of course, what we expect from the beginning since the numbers are sequences and not equivalence classes of sequences. Anyway, whatever appearance there may be of choice, it is clearly deceptive.

The whole idea of a shift from ordinary counting to a binary decimal computation at the first change in sign may seem unnatural at first. However, such phenomena seem inevitable in a sufficiently rich system.

It is an amusing exercise in arithmetic to add numbers in this form. Carrying exists as usual but since we deal with pluses and minuses and do not have zeros, an adjustment is necessary if we would otherwise obtain 0 in a place. Specifically, 0+ must be replaced by +−. Also, one must be aware of the dividing line where the shift from ordinary counting to the binary decimal computation occurs.

Finally, a suitable succinct way of expressing the result of arithmetical operations on dyadic fractions given in the above form may be useful in studying certain problems in the theory of surreal numbers. Although the dyadic fractions look like a drop in the ocean of surreal numbers, they form an important building block. As we shall see, for example in chapter seven, it is possible to ask questions which are non-trivial even for a set such as the dyadic fractions.

C REAL NUMBERS

At this stage we develop a theory which is somewhat analogous to that of Dedekind cuts. However, there is at least one important difference. All the objects and operations are already present as a subsystem of the surreal numbers. The analogy arises because of the theory in chapter two. First, roughly speaking, the fundamental existence theorem gives a well-defined element for every cut. Secondly, the canonical representation gives a natural cut associated with any element. This correspondence, of course, occurs at every stage, although the resemblance to Dedekind cuts is closer in some stages than others.

Definition. A real number is a surreal number a which is either of finite length or is of length ω and satisfies $(\forall n_0)(\exists n_1)(\exists n_2)\{[n_1 \geq n_0] \wedge [n_2 \geq n_0] \wedge [a(n_1) = +] \wedge [a(n_2) = -]\}$.

In other words, the definition simply requires that the terms of the sequence $a(n)$ of pluses and minuses do not eventually have constant signs. This is analogous to the situation for ordinary decimals where one might rule out an eventual sequence of nines to ensure that each number has a unique representation. In our case a sequence consisting eventually of pluses will be a surreal number outside the set of reals.

To show that the set of real numbers forms a field, it suffices to check the closure properties. However, it is convenient to have several lemmas in order to carry this through.

Note first that the distinction between surreal numbers of length ω which are real and those which are not can be expressed in terms of the canonical representation. If $a = F|G$ is the canonical representation, then a is real precisely when F and G are non-empty, F has no maximum and G has no minimum; e.g. if there is a last plus, then F has a maximal element. (It is clear from what we already know that this element is infinitesimally close to a but this issue will not be pursued now.)

The elements of $F \cup G$ have finite length so they are dyadic fractions.

Lemma 4.3. Let F and G be non-empty sets of dyadic fractions such that $F < G$, F has no maximum, and G has no minimum. Then $F|G$ is a real number.

Proof. $F|G$ exists by the fundamental existence theorem. Let $F|G = a$. By theorem 2.3 $\ell(a) \leq \omega$. It suffices to rule out the possibility that $\ell(a) = \omega$ and a has eventually constant sign. Suppose that the constant sign is $+$. (A similar argument applies if it is $-$.) The possibility that a consists exclusively of pluses is ruled out since $a < G$ and G is non-empty.

Now suppose $a(n_0) = -$ but $n > n_0 \to a(n) = +$. Let b be the initial segment of a of length n_0. Then $b > a$. By theorem 2.9, $(\exists c \varepsilon G)(c \leq b)$ since b is an upper element in the canonical representation of b. Since G has no minimum $(\exists d \varepsilon G)(d < c)$. Since b, c, and d are all dyadic fractions we can choose m so that $d \leq b - \frac{1}{2^m}$.

Let c be an initial segment of a of length n for some $n > n_0$. Then c is a lower element. Also, $c-b$ has the form $-\frac{1}{2^r} + \frac{1}{2^{r+1}} \cdots + \frac{1}{2^{r+s}} = -\frac{1}{2^{r+s}}$ for some r and s. (Actually $s = n-n_0-1$.) In any case, for n sufficiently high, $c > b - \frac{1}{2^m}$. Hence $a > c > b - \frac{1}{2^m}$. Also $a < d \leq b - \frac{1}{2^m}$. Thus we have the desired contradiction.

Lemma 4.4. Let $a = F|G$. Suppose that $(\forall x \varepsilon F)(\exists$ a positive dyadic fraction $r)(\exists y)(y \geq x+r \wedge y \varepsilon F)$ and $(\forall x \varepsilon G)(\exists$ a positive dyadic fraction $r)(\exists y)(y \leq x-r \wedge y \varepsilon G)$.

Also let $F' < a < G'$. Suppose that $(\forall$ positive dyadic $r)(\exists x \varepsilon F')(\exists y \varepsilon G')(y-x \leq r)$.

Then $a = F'|G'$.

Proof. It is enough to check cofinality. We do this for F'. The case for G' is similar. Suppose $x \varepsilon F$. Choose y and r such that r is positive dyadic, $y \geq x+r$ and $y \varepsilon F$. For that same r choose $w \varepsilon F'$ and $z \varepsilon G'$ such that $z - w \leq r$. Since $F < a$ we have $a > y \geq x+r$. Since $G' > a$ we have $w \geq z-r > a-r \geq x$. Since $w \varepsilon F'$ this proves cofinality.

Note that the hypothesis does not require any of the sets to consist only of dyadic fractions.

Lemma 4.5. There are an infinite number of dyadic fractions between any two distinct real numbers a and b.

Proof. It clearly suffices to obtain one dyadic fraction. If neither a nor b is dyadic then the common segment works. More generally, if neither a nor b is an initial segment of the other, we can use the common segment. Suppose a is a proper initial segment of b. (a is necessarily dyadic.) If b is dyadic the result is trivial. Otherwise consider the canonical representation $F|G$ of b. $a \in F \cup G$. Since F has no maximum and G no minimum we obtain a dyadic fraction between a and b whether $a \in F$ or $a \in G$.

Remark. Note that if a has a last plus and b is the initial segment of a which stops just before the last plus, then there is no dyadic fraction between a and b. Thus the requirement in the lemma that the numbers be real is essential.

Lemma 4.6. Let $a = F|G$ be the canonical representation of a real number a which is not a dyadic fraction. Then for all positive dyadic r there exist $b \in F$, $c \in G$ such that $c-b \leq r$.

Proof. Since there is no last + and no last - in a, then for all n there are elements $b \in F$, $c \in G$ which agree in the first n terms. Thus c-b is bounded above by an expression of the form $\frac{1}{2^s} + \frac{1}{2^{s+1}} \cdots + \frac{1}{2^{s+1}} \leq \frac{1}{2^{s-1}}$. Since s can be made arbitrarily large by a suitable choice of n this proves the lemma.

Note that it is easy to see that the requirement that a be real can be relaxed but this is of no special concern.

We are now ready to check the closure properties.

Addition. Let a and b be real numbers. If both are dyadic fractions then so is the sum. Suppose a is a dyadic fraction and b is not. Let $a = F|G$ and $b = F'|G'$ be the canonical representations. Then $a+b = \{a'+b, a+b'\}|\{a''+b, a+b''\}$. We claim that numbers of the form $a+b'$ are cofinal on the left. Consider a number of the form $a'+b$. Since a, a' are dyadic fractions and $a > a'$, $a-a'$ is a positive dyadic fraction. By lemma 4.6, $(\exists b' \in F')(\exists b'' \in G')(b''-b' \leq a-a')$. Then $a+b' \geq a'+b'' > a'+b$. Similarly, we can see that numbers of the form $a+b''$ are cofinal on the right. By the cofinality theorem $a+b = \{a+b'\}|\{a+b''\}$. Note that $a+b'$ and $a+b''$ are dyadic fractions. Also since $F' = \{b'\}$ has no maximum, neither does $\{a+b'\}$. Similarly

$\{a+b''\}$ has no minimum. By lemma 4.3, $a+b$ is a real number.

Now suppose neither a nor b is a dyadic fraction. Again let $a+b = \{a'+b,\ a+b'\}|\{a''+b,\ a+b''\}$. As before it is clear that the left elements have no maximum and the right elements no minimum. More specifically, since the numbers a', a'', b', b'' are all dyadic the above representation of $a+b$ satisfies the first condition of lemma 4.4. Now let $F' = \{a'+b'\}$ and $G' = \{a''+b''\}$. Then $F' < a+b < G'$. Also lemma 4.6 together with the usual "$\frac{\varepsilon}{2}$ argument" shows that F' and G' satisfy the other conditions of lemma 4.4. Hence $a = F'|G'$. Finally, F and G' satisfy the hypothesis of lemma 4.3 so a is real.

0 and 1 are clearly real. The additive inverse of a real is real since the definition is symmetric in pluses and minuses.

Multiplication. Let a and b be real numbers. If both are dyadic fractions then so is the product. Now let a be a real number which is not a dyadic fraction. There is no restriction on b. A typical element in the representation of ab is $c = ab - (a-a^o)(b-b^o)$. [Recall that, e.g., a^o is either of the form a' or a''.] Since a is real we can choose $a_1{}^o$ which is on the same side of a as a^o and also closer to a (we are using a unified argument). If we use the same b^o we obtain the element $c_1 = ab -(a-a_1{}^o)(b-b^o)$. Then $c_1-c = (a_1{}^o-a^o)(b-b^o)$. Now by applying lemma 4.5 to an arbitrary positive real r and 0 we obtain a positive dyadic fraction d satisfying $d < r$. For a product of positive reals r_1 and r_2 we obtain in this manner the positive dyadic fraction d_1d_2 satisfying $d_1d_2 < r_1r_2$. Thus we have what we need to show that the first condition of lemma 4.4 is satisfied for the representation of ab. The choice of F' and G' would depend on the signs of a and b; however it suffices to assume that $a,b > 0$.

Let $F' = \{a_1b_1\colon a_1 \text{ is dyadic} \wedge 0 \leq a_1 < a \quad b_1 \text{ is dyadic} \wedge 0 \leq b_1 < b\}$ and $G = \{a_1b_1\colon a_1 \text{ is dyadic} \wedge a < a_1 \wedge b_1 \text{ is dyadic} \wedge b < b_1\}$. (This formulation is convenient since we can obtain the required conclusion if b is a dyadic fraction even though lemma 4.6 does not apply in that case.) For arbitrary dyadic r we can choose a_1, a_2, b_1, b_2 such that $b_2-b_1 \leq r$ and $a_2-a_1 \leq r$. The usual argument for proving that the limit of a product is the product of the limits shows that the other conditions of lemma 4.4 are satisfied. Note that we use the fact that for every real number a there exists an integer n such that $|a| \leq n$. This is clear since in the canonical representation of a as $F|G$ both F and G are non-empty. (Incidentally, for the surreal number of length ω which consists solely of pluses the argument would break down.)

Hence by lemma 4.4 $ab = F'|G'$. Again, as in the case of addition, F' and G' satisfy the hypothesis of lemma 4.3, so that ab is real.

Note that during the proofs of closure we obtained nice representations of sums and products of reals with the help of lemma 4.4. Since such intuitive-looking representations fail in general, it is interesting that they work in the special case of real numbers.

Reciprocals. It is best to ignore the previous construction of reciprocals. It suffices to consider the case where a is a real number larger than 0. Let $F = \{d: d$ is a dyadic fraction $\wedge\, da < 1\}$ and $G = (d: d$ is a dyadic fraction $\wedge\, da > 1)\}$. Clearly $F < G$, F and G are non-empty and $0 \in F$. Also by lemma 4.5 we can find m so that $a > \frac{1}{2^m}$. Hence $2^m a > 1$.

We now show that F has no maximum. Suppose d is a dyadic fraction satisfying $da < 1$. Then $1 - da$ is a real number because of the closure properties, and $1 - da > 0$. By lemma 4.5 one can find m so that $1 - da > \frac{1}{2^m}$. Now we already noted earlier that a is bounded above by an integer which we may just as well call 2^n, i.e. $a < 2^n$. Hence $\frac{1}{2^{m+n}}\ a < \frac{1}{2^m} < 1 - da$. Thus $\left(d + \frac{1}{2^{m+n}}\right) a < 1$. Similarly G has no minimum.

By lemma 4.3 $b = F|G$ is a real number. By closure of multiplication ba is a real number. Now let n be arbitrary. We would like to show that there exists an element c in G such that $ca \leq 1 + \frac{1}{2^n}$. Choose m such that $a < 2^m$.

We now consider the set $P = \{r: (\frac{r}{2^{m+n}})\ a > 1\}$. As a non-empty set of positive integers, P has a least element s. Then by definition $\frac{s}{2^{m+n}} \in G$. Furthermore $(\frac{s-1}{2^{m+n}})\ a \leq 1$. Hence $(\frac{s}{2^{m+n}})\ a \leq 1 + \frac{1}{2^n}$. Similarly there exists an element $c' \in F$ such that $c'a \geq 1 - \frac{1}{2^n}$. Hence $1 - \frac{1}{2^n} \leq c'a < ba \leq ca \leq 1 + \frac{1}{2^n}$. This shows that

$|ba-1| < r$ for any positive dyadic fraction r. It follows from lemma 4.5 that $ba = 1$. This completes the proof.

Note that reciprocals of dyadic fractions are not necessarily dyadic fractions. Thus a naive attempt consisting of expressing a as $F|G$ and using something like $\frac{1}{G}|\frac{1}{F}$ to obtain the reciprocal would cause dfficulties. Our proof circumvents that problem.

Now we know that the real numbers form a field containing the dyadic fractions. They therefore contain all the rational numbers.

Finally, we would like to prove the l.u.b. property, i.e. that every bounded non-empty set of reals has an l.u.b. within the set of reals. It's important to note that in our development the real numbers form a proper subset of the system we are studying so that care is needed in stating the l.u.b. property. In fact, as a special case of theorem 2.1 we see that any set which has no maximum has no l.u.b. in the class of surreal numbers.

Incidentally, it is immediate from lemma 4.5 that every real number is the l.u.b. of all dyadic fractions below it but this is not what we really need. Since the "ordinary" real numbers can be characterized as an ordered field with the l.u.b. property, the latter is all that remains to be proved.

Suppose H is a non-empty set of real numbers bounded above. Let G be the set of upper bounds which are dyadic fractions and let F be the set of all other dyadic fractions. F and G are clearly non-empty. By lemma 4.5, F has no maximum. If G has a minimum b, then b is already an l.u.b. to H and we are done. (Again, lemma 4.5 is used since it guarantees that a least upper bound among the set of dyadic fractions is automatically a least upper bound in the set of all real numbers. So suppose G has no minimum. Then by lemma 4.3 $r = F|G$ is a real number. We now show that l.u.b. $H = r$. First, r is an upper bound to H. Otherwise, $(\exists a \in H)(r<a)$. Let d be a dyadic fraction satisfying $r < d < a$. (We are getting our money's worth out of lemma 4.5!) Since $d < a \in H$, $d \in F$. However $r < d$, contradicting the fact that $F < r$. Finally, suppose $s < r$ and s is an upper bound to H. Let $s < d < r$ for some dyadic d. Then d is also an upper bound to H. Hence $d \in G$. But $d < r$. This contradicts the fact that $r < G$.

We have finally shown the following.

Theorem 4.3. The real numbers form an ordered field with the l.u.b. property (i.e they are essentially the same as the "reals" defined in more traditional ways.)

We close this section with several remarks. Let a be a real number which is not a dyadic fraction, thus a has length ω. If $a = F|G$ is the canonical representation, we know that $F < a < G$ and that the elements of F form an increasing sequence and that the elements of G form a decreasing sequence. Since F has no maximum and G no minimum, both sequences are infinite. It is clear from this and from lemma 4.6 that if a_n is the initial segment of A of length n then $\lim_{n\to\infty} a_n = a$. Of course, in the definition of a limit it makes no difference whether ε is taken to be real, rational, or dyadic; however, we certainly cannot use general surreal numbers.

Finally, we compare our definition with the one used in [1] which makes no reference to sequences. It is convenient for motivation to consider a third definition which is, roughly speaking, intermediate in spirit between the two definitions.

Theorem 4.4. The following three conditions are equivalent.

(a) a is a real number (by our definition);

(b) For some integer n, $-n < x < n$ and a has no initial segment a_α such that $|a-a_\alpha|$ is infinitesimal;

(c) For some integer n, $-n < x < n$; and $a = \{a-1, a-\frac{1}{2}, a-\frac{1}{3}\cdots\}|\{a+1, a+\frac{1}{2}, a+\frac{1}{3}\}$ (the definition in [1]).

Remarks. An element a is infinitesimal if it is nonzero but for every positive rational r, $|a| < r$. It is clear, for example, from what we already know that the element a of length ω which begins with $+$ and then after that consists only of $-$'s is an infinitesimal. The heuristic idea in definition (c) is that of the possibility of writing a in the form $F|G$ without forcing either F or G to be "too close" to a. Of course, by the cofinality theorem, if $a = F|G$ then one still gets a if F and G are both enlarged to include elements closer to a. The challenge lies in the opposite direction. How far away can F and G be from a and still have $a = F|G$? As a rough rule of thumb, the larger the length of a, the closer F and G must be to a.

Proof. (a) ⇒ (b). Since all initial segments are dyadic fractions, this is clear.

Not (a) ⇒ not (b). (This contra-positive notation seems natural here even if it may be unusual.)

Suppose a is not real. Then $\ell(a) \geq \omega$. If for all n less than ω, a has a fixed sign then the condition $-n < a < n$ fails, as is clear from the ordering, so that case is clear. Now let a_ω be the initial segment of a of length ω. Assume first that a_ω consists eventually of pluses only. (A similar argument will apply if ω consists eventually of minuses only.) Then we can use an argument which is essentially the same as the one used in the proof of lemma 4.3. We let a_n be the initial segment of a_ω obtained by stopping just before the last minus. (We already ruled out the case where a_ω contains no minuses.) Then $a_n > a$. For all positive rational r we can find m sufficiently high such that the initial segment of a_m of a of length m satisfies $a_m < a$ and $a_n - a_m < r$. As in the proof referred to above, we use a computation of the form $-\frac{1}{2^s} + \frac{1}{2^{s+1}} + \frac{1}{2^{s+2}} + \ldots$. Hence $a_n - a < r$ for all r. Since n is fixed, we have $|a_n - a|$ is infinitesimal. Note that this part of the argument is independent of whether a is the same as a_ω or not.

Now assume that a_ω does not eventually have constant sign. Since a is not real, $a \neq a_\omega$., i.e. a_ω is a proper initial segment of a. Let r be an arbitrary positive rational. Suppose $a_\omega = F|G$. Then by lemma 4.6 there exist $b \in F$, $c \in G$ such that $c-b \leq r$. Now $b < a_\omega < c$. It is also clear that $b < a < c$ by the lexicographical ordering. (In fact, b and c occur in the canonical representation of a as well as a_ω.) Hence $|a_\omega - a| < r$. Therefore $|a_\omega - a|$ is infinitesimal.

(b) ⇒ (c). Express a canonically as $F|G$. Then the conclusion is immediate by the cofinality theorem. Given $a' \in F$, then $a - a'$ is not infinitesimal. Choose n so that $a - a' > \frac{1}{n}$. Then $a - \frac{1}{n} > a'$. A similar argument applies to $a' \in G$.

(c) ⇒ (b). Let $a = \{a-1, a - \frac{1}{2}, a - \frac{1}{3}, \ldots\} | \{a+1, a + \frac{1}{2}, a + \frac{1}{3}, \ldots\}$ and let $a = F|G$ canonically. Then by the inverse cofinality theorem the set $\{a-1, a - \frac{1}{2}, a - \frac{1}{3} \ldots\}$ is cofinal in F. Let $a' \in F$. Then for some n, $a - \frac{1}{n} \geq a'$, i.e. $a - a' \geq \frac{1}{n}$. Therefore $a - a'$ is not infinitesimal. A similar argument applies to G.

In line with what was said earlier, note as a point of caution that the theorem does not rule out the possibility that a is real of the form $F|G$ with elements in F or G infinitesimally close to a. This is ruled out only for the canonical representation.

D ORDINALS

Our next task is to show that the surreal numbers contain the ordinals. Once this is done it will be legitimate to deal with expressions of the form $\omega-1$ or $\frac{1}{2}\omega$. First, of course, we have to make precise what we mean by the statement that the ordinals are a subclass of the surreal numbers. Recall that ordinary addition and multiplication on the ordinals are not commutative. However, there do exist natural commutative operations on the ordinals that have been considered in the literature and they do correspond to the operations we defined on the surreal numbers.

We identify the ordinal α with the sequence a_α of length α such that $(\forall n<\alpha)[a(n) = +]$.

First note that by theorem 4.1 this is consistent with the situation for positive integers. Also it is immediate from the lexicographical order that $\alpha < \beta \Rightarrow a_\alpha < a_\beta$. Furthermore, the canonical representation of a_α is $\{a_\beta: \beta<\alpha\}|\phi$. Again, if H is a set of ordinals then $\{a_H\}|\phi = a_\alpha$ where α is the least ordinal such that $\alpha > H$. If H has a maximum β then $\alpha = \beta+1$. If H has no maximum then α = l.u.b. H. α is called the sequent of H and denoted by seq H in the literature. In summary, as far as the order properties are concerned, the identification is quite reasonable. Thus for convenience of notation we use α instead of a_α. So the ordinals are the sequences which consist only of pluses. (Incidentally, note that our definition of cofinality [see p. 9] is consistent with the usual one for ordinals.)

We now show that addition and multiplication correspond to what is commonly called natural addition and natural multiplication. They are obtained by taking the usual expansion in powers of ω and operating as if they are ordinary polynomials (i.e. no absorption). In order to state the next theorem precisely we tentatively use + for ordinary addition, $\oplus$ for natural addition and $\dot{+}$ for surreal addition.

Theorem 4.5. For any ordinals α and β, $\alpha\dot{+}\beta = \alpha\oplus\beta$.

Proof. We use induction as usual. In view of our earlier remarks $\alpha\dot{+}\beta = \operatorname{seq}_{\gamma<\beta,\delta<\alpha} (\alpha\dot{+}\gamma, \delta\dot{+}\beta) = \operatorname{seq}_{\gamma<\beta,\delta<\alpha} (\alpha\oplus\gamma, \delta\oplus\beta)$ by the inductive hypothesis. The problem is now reduced to an elementary exercise in the ordinary theory of the ordinals. Specifically, we may express α and β respectively in the forms $\alpha'+\omega^r$ and $\beta'+\omega^s$ (r or s may be 0). Then typical lower elements are $(\alpha'+\omega^r) \oplus (\beta'+\gamma)$ with $\gamma < \omega^s$ and $(\alpha'+\delta) \oplus (\beta'+\omega^s)$ with $\delta < \omega^r$. Without loss of generality suppose $r \geq s$. Then the set $(\alpha'+\omega^r) + (\beta'+\gamma)$ with $\gamma < \omega^s$ is cofinal and we clearly obtain $(\alpha'+\omega^r) \oplus (\beta'+\omega^s)$ as the sequent which is what we want.

In order to state the next theorem precisely we tentatively use $\otimes$ for natural multiplication and $\dot{\times}$ for surreal multiplication.

Theorem 4.6. For any ordinals α and β $\alpha\dot{\times}\beta = \alpha\otimes\beta$.

Proof. Again we use induction. Now every ordinal has the form $\sum_{i=1}^{k} \omega^{r_i} n_i$ with $r_i > r_{i+1}$ for all i and all n_i integers. This may also be written in the form $\sum_{i=1}^{k} \omega^{r_i}$ with $r_i \geq r_{i+1}$ by breaking up all $\omega^{r_i} n_i$ for which $n_i > 1$. The sum is also a natural sum and therefore by theorem 4.5 a surreal sum. Suppose $\alpha = \sum \omega^{r_i}$ and $\beta = \sum \omega^{s_i}$. By the distributive law for surreal numbers we obtain $\alpha\dot{\times}\beta = \sum_{i,j} \omega^{r_i} \dot{\times} \omega^{s_j}$.
If at least one of α and β is not a power of ω, then we can use the inductive hypothesis to obtain

$$\alpha\dot{\times}\beta = \sum_{i,j} \omega^{r_i} \otimes \omega^{s_j} = \alpha\otimes\beta$$

by the distributive law for natural multiplication over natural addition.

Now suppose that both a and b are powers of ω. Let $\alpha = \omega^r$ and $b = \omega^s$. Then by definition $\alpha\dot{\times}\beta = \{(\omega^r\dot{\times}\delta) \dot{+} (\gamma\dot{\times}w^s) - (\delta\dot{\times}\gamma)\}|\phi$ where $\delta < \omega^s$ and $\gamma < \omega^r$. Note that unlike in the case of addition, we are stuck with the surreal operation - which does not correspond to an operation on ordinals.) First, since there are no upper elements, by theorem 2.2 $\alpha\dot{\times}\beta$ is an ordinal.

The typical lower element is certainly not larger than $(\omega^{r}\dot{\times}\delta) \dot{+} (\gamma\dot{\times}w^{s})$ which by the inductive hypothesis and theorem 4.5 is $(\omega^{r}\otimes\delta) + (\gamma\otimes w^{s})$. Since $\delta < \omega^{s}$ and $\gamma < \omega^{r}$. This is clearly less than ω^{r+s}. Hence $\omega^{r\oplus s}$ satisfies the betweenness property in the definition of $\alpha\otimes\beta$. Hence $\ell(\alpha\dot{\times}\beta) \leq \omega^{r\oplus s}$ and since $\alpha\dot{\times}\beta$ is an ordinal, $\alpha\dot{\times}\beta \leq \omega^{r\oplus s}$.

Now if $r' < r$ and n is an arbitrary positive integer we have $\alpha\dot{\times}\beta > \omega^{r'}n\dot{\times}\omega^{s} = \omega^{r'}n\otimes\omega^{s}$ by the inductive hypothesis. Furthermore this is $\omega^{r'\oplus s}n$. Similarly if $s' < s$ we have $\alpha\dot{\times}\beta > \omega^{r\oplus s'}n$. We have already noted during our proof of theorem 4.5 that seq($r'\oplus s$, $r\oplus s'$) = $r\oplus s$. By the basic properties of the expansion of ordinals in powers of ω, it follows that $\text{seq}(\omega^{r's}n, \omega^{r+s'}n) = \omega^{r+s}$. Hence $\alpha\dot{\times}\beta \geq \omega^{r+s}$. So finally $\alpha\dot{\times}\beta = \omega^{r\oplus s} = \alpha\otimes\beta$.

In view of theorems 4.5 and 4.6 we no longer need symbols such as $\dot{+}$ and $\dot{\times}$. For convenience we will even drop the symbols $\oplus$ and $\otimes$ since it will be clear from the context whether natural or ordinary operations are intended. In fact, as a rule of thumb, in discussing elements we use natural operations which, as we have just shown, are the same as the surreal operations. On the other hand, when discussing lengths and juxtaposition of sequences, the ordinary operations are appropriate. This issue involving choice of notation will occur in other places as well. In general, readability and reliance on context will take precedence over a picayune attitude which complicates notation unnecessarily.

We now know the exciting fact that the surreals form a field containing both the reals and ordinals. So, for example, elements such as $\omega-1$ and $\frac{1}{2}\omega$ have meaning. Incidentally, our $\omega-1$ has nothing to do with a meaning used in the literature, namely $\omega-1 = \omega$, since $1+\omega = \omega$ where $+$ stands for ordinal addition. We shall compute the sign sequence of several such strange elements. These are all special cases of the general theory in the next chapter. However, it is worthwhile to see some elementary concrete examples before getting involved with a general representation theorem. In fact, there is some pedagogical value for the reader to experiment with other elementary examples before coping with a more general and abstract situation.

We begin with $\omega-1$ which is about the simplest looking "exotic" element. This is $\omega+(-1)$. The canonical representation of ω

is $\{n\}|\phi$ where n stands for an arbitrary non-negative integer and $-1 = \phi|\{0\}$. Hence using the definition of addition $\omega-1 = \{n-1\}|\{\omega+0\}$. By cofinality theorem b (theorem 2.7), the left-hand side may be replaced by $\{n\}$; hence we have $\{n\}|\{\omega\}$ which we can see immediately from the definition is the sequence of length $\omega+1$ which consists of ω pluses followed by a minus.

We make several remarks before continuing with other examples. It is often convenient to use theorem 2.7 to simplify expressions of the form $F|G$. Even though there is an option of using theorem 2.6, theorem 2.7 has the convenience of avoiding specific reference to $a = F|G$ but referring only to F and G themselves. This helps to streamline a computation; in fact, in future we shall use theorem 2.7 freely without quoting it if its use is obvious (just as in elementary algebra one does not bother to quote the distributive law every time it it is used!). For example, an expression such as $\{n + \frac{5}{2}\}|\{\omega - \frac{1}{2m}\}$ where m and n run through all positive integers can be replaced by $\{n\}|\{\omega - \frac{1}{m}\}$.

Note that the last part of the computation can be done in two ways. One can directly use the definition and see that any x satisfying $n < x < \omega$ for all n must necessarily begin with the sequence of length $\omega+1$ referred to above. One can also take a good guess at the answer and note that $\{n\}|\{\omega\}$ happens to be the canonical representation. This is worth emphasizing, because in more complicated situations we often have a representation which is cofinal in a canonical representation. In these cases it is much easier to use the second method.

Recall that by the uniformity theorems, the computations may be performed using any representations; thus we may choose ones which appear to be most tractable. For example, for any real number a we may use the representation $a = F|G$ where F is the set of all dyadic fractions below a and G is the set of all dyadic fractions above a. This is a unified formula which applies whether or not a is a dyadic fraction. If a is not a dyadic fraction, this is mutually cofinal with the standard representation, but it definitely is not mutually cofinal if a is a dyadic fraction. I.e. in the latter case the representation cannot be justified by theorem 2.7, although theorem 2.6 clearly applies.

As obvious as these above remarks may be, they are worthwhile to state and get out of the way once and for all, since it would be clumsy to make them in the middle of a proof. With this in mind, we can now present computations and proofs more efficiently, i.e. using convenient representations and simplifications without explicitly stating the obvious justification.

Now consider $\omega-2$. This is $\omega+(-2) = \{n\}|\phi+\phi|\{-1\} = \{n-2\}|\{\omega-1\} = \{n\}|\{\omega-1\}$ which is the sequence of length $\omega+2$ consisting of ω pluses followed by two minuses.

Using induction we can compute $\omega-(m+1)$ for a positive integer n. In fact, $\omega-(m+1) = \{n\}|\phi+\phi|\{-m\} = \{n-m-1\}|\{\omega-m\}$. (To avoid confusion note that m is fixed although n varies.) This is $\{n\}|\{\omega-m\}$ which, using the obvious inductive hypothesis, gives the sequence of length $\omega+m+1$ consisting of ω pluses followed by $m+1$ minuses.

It is natural now to ask what ω pluses followed by ω minuses represents. The pattern suggests naively that it may be $\omega-\omega$ but this is obviously impossible. That element is, of course, still infinite, i.e. above every positive integer because of the lexicographical order. In fact, no sequence of minuses no matter how large can undo the effect of the first ω pluses. We shall return to the above example shortly.

The computation for $\omega-m$ works for any limit ordinal as well as ω. For example $(\omega^2+6\omega) - 3$ is the sequence of $\omega^2 + 6\omega$ pluses followed by 3 minuses. One can also apply the computation to non-limit ordinals to check the consistency of what we have already done. For example $(\omega+6) - 1$ is simply $\omega+5$. The computation will thus not give you the sequence of $\omega+6$ pluses followed by a minus if it is done correctly. The step to beware of is the following. For a limit ordinal α we can simplify a lower set such as $\{\beta-1:\beta<\alpha\}$ by replacing it by $\{\beta: \beta<\alpha\}$. This is, of course, not valid for non-limit ordinals, since the former set is not cofinal in the latter.

We now compute $\omega + \frac{1}{2}$. This is $\{n\}|\phi + \{0\}|\{1\} = \{n+\frac{1}{2}, \omega+0\}|\{\omega+1\} = \{\omega\}|\{\omega+1\}$. This is the sequence of $\omega+1$ pluses followed by a minus. In line with our earlier computations, this example illustrates the contrast between the cases where α is a limit ordinal and where α is a non-limit ordinal in the value of a sequence with α pluses followed by a minus.

In the same manner induction can be used to evaluate $\omega+r$ for any positive real number r. $\omega+r = \{n\}|\phi + F|G$ where $F|G$ is the canonical representation of r. This is $\{\omega+F, n+r\}|\{\omega+G\}$. Since $0 \in F$, the left-hand side may be replaced by $\omega+F$ and we obtain $\{\omega+F\}|\{\omega+G\}$. By the lexicographical order and the inductive hypothesis, this is the sequence with ω pluses followed by the sequence for r. Incidentally, this reasoning with the lexicographical order and juxtaposition is similar to what has already been used back in the proof of theorem 2.1 and represents an important skill to facilitate computation.

This argument works just as well when r is negative as long as r is not an integer, in which case F is empty. The latter case has in fact been done earlier and the general conclusion is still valid although $\omega+F$ can no longer be used as a lower set.

As before, similar results apply if ω is replaced by other limit ordinals.

We now consider $\frac{1}{2}\omega$. This is $(\{0\}|\{1\})\times(\{n\}|\phi)$. By the definition of multiplication this is $\{\frac{1}{2}n+\omega.0-n.0\}|\{\frac{1}{2}n+\omega.1-n.1\} = \{\frac{1}{2}n\}|\{\omega - \frac{1}{2}n\} = \{n\}|\{\omega-n\}$. Using the earlier result for $\omega-n$, this is the sequence of length $\omega.2$ beginning with ω pluses and followed by ω minuses. This answers an earlier question which was left open.

It is instructive to see another proof. Let a be the sequence of ω pluses followed by ω minuses. Then $a = \{n\}|\{\omega-n\}$. Hence $2a = a+a = \{n+a\}|\{\omega-n+a\}$. We now prove that this is ω by cofinality. We know that for all positive integers n, $n < a < \omega-n$. Hence $n+a < \omega$. Since $a > n$, it follows that $\omega < \omega-n+a$. So ω satisfies the betweenness condition. Since $\omega = \{n\}|\phi$ and $n+a \geq n$, the cofinality condition is satisfied; hence $2a = \{n+a\}|\{\omega-n+a\} = \omega$.

The first proof involves more computation but is more routine. The second method requires a good guess at the answer and being able to use cofinality in spite of a limited knowledge of a.

$\frac{1}{2}\omega-1$ and more generally $\frac{1}{2}\omega+r$ for arbitrary real r can be handled similarly to $\omega-1$ and $\omega+r$. In fact, the sequence for $\frac{1}{2}\omega+r$ is the sequence for $\frac{1}{2}\omega$ followed by the sequence for r. Such simple results make the subject quite tractable. However, the reader is warned that crude juxtaposition does not work in every case.

Several more remarks are worthwhile to mention at this point. First, we take for granted obvious inequalities involving infinite or infinitesimal elements (e.g., for all positive real r and integers n, $\omega r-n$ is positive infinite) and apply them to obtain information about cofinality. Secondly, recall the definition of multiplication and consider $a = A'|A''$ and $b = B'|B''$. Then ab is $\{a'b+ab'-a'b', a''b+ab''-a''b''\}|\{a'b+ab''-a'b'', a''b+ab'-a''b'\}$. It is often convenient to think of the lower sums in the form $a'b + (a-a')b'$ and $a''b - (a''-a)b''$ and the first upper sum in the form $a'b + (a-a')b''$ or $ab'' - a'(b''-b)$. In spite of the triviality of the algebra, a suitable form supplies a considerable gain in intuition.

We now compute $\frac{3}{4}\omega$. $\frac{3}{4} = (+-+) = \{\frac{1}{2}\}|\{1\}$ by our work on dyadic fractions. Hence $\frac{3}{4}\omega = (\{\frac{1}{2}\}|\{1\})\times(\{n\}|\phi) = \{\frac{1}{2}\omega + (\frac{3}{4} - \frac{1}{2})n\}|\{1.\omega-(1 - \frac{3}{4})n\}$. Note the use of the above remarks. Also note that a typical lower term may alternatively be written as $\frac{3}{4}n + \frac{1}{2}(\omega-n)$, but the form we have exhibits the order of magnitude more clearly. In fact, it is immediate by cofinality that $\frac{3}{4}\omega = \{\frac{1}{2}\omega+n\}|\{\omega-n\}$. If we accept the juxtaposition results for $\frac{1}{2}\omega+n$ we see that this is the sequence of length $\omega.3$ consisting of ω pluses followed by ω minuses and then ω pluses.

By a similar process we can use a double induction to obtain the sequence for ωr, one for n in $\omega r\pm n$ and one for r. It turns out to be just like the sequence for r except that each sign is repeated ω times. The earlier remarks on multiplication which are petty in an individual numerical example gain in value as we generalize.

Now we consider $\omega^2-\omega$. It is easy to see that in this case the induction procedure for ω^2-n extends even to $\omega^2-\omega$ so we obtain that $\omega^2-\omega$ consists of ω^2 pluses followed by ω minuses. The key point is the elementary fact about ordinals which says that $\alpha < \omega^2 \rightarrow \alpha+\omega < \omega^2$. The reader can experiment with obvious generalizations in this direction and see that there are no further dramatic surprises.

Before studying the ultimate representation theory, it is worthwhile to see what happens in the other direction, i.e. with expressions such as $\frac{1}{\omega}$. We prefer to ignore our previous construction of reciprocals just as we did in our study of the real numbers. Our primary interest in the latter construction was, of course, in the issue of existence since it is usually not convenient to use such an inductive construction for computation even if it is possible.

Instead, we use the good guess approach. Let ε be the sequence of length ω which consists of a single plus followed by ω minuses. This is clearly a positive infinitesimal. In fact, it is immediate that it is the unique positive infinitesimal of length ω; hence it may be regarded as the canonical infinitesimal. Heuristically, this is a reasonable candidate for being the reciprocal of the canonical infinitely large number ω. We now prove this fact. Note that the canonical representation of ε is $0|\{\frac{1}{2^n}\}$. Hence $\varepsilon\omega = (0|\{\frac{1}{2^n}\})\times(\{m\}|\phi) = (0|\{\frac{1}{n}\})\times(\{m\}|\phi) = \{\varepsilon m+0\omega-0\omega\}|\{\varepsilon m + \frac{1}{n}\omega - \frac{1}{n}m\} = \{\varepsilon m\}|\{\varepsilon m + \frac{1}{n}\omega - \frac{1}{n}m\}$. Now $1 = \{0\}|\phi$. We now check the conditions of the cofinality theorem. Since ε is infinitesimal, a typical lower element which is εm is less than 1. A typical upper element $\frac{1}{n}\omega - \frac{1}{n}m + \varepsilon m$ is clearly infinite. Hence 1 satisfies the betweenness condition. The cofinality part requires minimal work. Regardless of m, $\varepsilon m \geq 0$. Hence we do obtain 1 as required.

Note that in spite of the existence of infinitesimals there is no connection with nonstandard analysis. So far we have not had any transfer principle. In fact, no model-theoretic ideas of any kind have played a role.

Let us now consider $r + \varepsilon$ where r is a real number. First, let r be a dyadic fraction. Then r has the form $\{s\}|\{t\}$ for suitable unit sets where s and t are also dyadic. Then $r + \varepsilon = \{s\}|\{t\} + \{0\}|\{\frac{1}{n}\} = \{r+0, s+\varepsilon\}|\{r + \frac{1}{n}, t+\varepsilon\} = \{r\}|\{r + \frac{1}{n}\} = \{r\}|G$ where G is the set of all dyadic fractions above r. It is immediate from the definition that this is the juxtaposition of the sequence for for r with the sequence for $\frac{1}{\omega}$. Note that this argument applies to all dyadics including integers since by cofinality we can always insert an s or t. Note also that we can now fill a gap which remained since our discussion of the real numbers, since we see from the above that all sequences whose terms are eventually minus are obtained this way. Thus all sequences of length ω are either real, $\pm\omega$, of the form $r + \frac{1}{\omega}$ for dyadic r, or of the form $r - \frac{1}{\omega}$ for dyadic r.

Now let r be non-dyadic. Let $r = R'|R''$. Then $r+\varepsilon = R'|R'' + \{0\}|\{\frac{1}{n}\} = \{r+0, r'+\varepsilon\}|\{r + \frac{1}{n}, r''+\varepsilon\} = \{r\}|\{r + \frac{1}{n}\} = r|G$ where G is the set of all reals above r. (G may be taken to be the set of dyadics instead. It makes no difference.) It is immediate from the definition that we now obtain the sequence for r followed by a single plus. Thus we have a sequence of length $\omega+1$. We have here a counter-example to naive juxtaposition. In fact, the final "poor" plus is worth only ε! Note that in both cases we are juxtaposing a sequence to the sequence for r but the sequence added on depends on whether r is dyadic or not. This brings some subtlety to the subject.

Next, we compute 2ε. This is $\{0\}|\{\frac{1}{n}\} + \{0\}|\{\frac{1}{n}\} = \{0+\varepsilon\}|\{\frac{1}{n}+\varepsilon\} = \{\varepsilon\}|\{\frac{1}{n}+\varepsilon\} = \{\varepsilon\}|\{\frac{1}{n}\}$ by mutual cofinality. Hence 2ε is the sequence for ε followed by a plus. Note the contrast between this case and that of 2ω. Again the final plus has value only ε.

Consider $\frac{1}{2}\varepsilon$. This is $(\{0\}|\{1\})\times(\{0\}|\{\frac{1}{n}\})$ which is $\{0,\frac{1}{2n} + \varepsilon - \frac{1}{n}\}|\{\frac{1}{2n},\varepsilon\} = \{0\}|\{\varepsilon\}$. This is the sequence of length $\omega+1$ which is that of ε followed by a minus. We can say that the value of the last sign is only $\frac{1}{2}\varepsilon$. The inflation rate is growing rapidly! Again note the contrast between this case and that of $\frac{1}{2}\omega$ where ω minus signs took care of the $\frac{1}{2}$.

Next we consider $r\varepsilon$ for a typical positive real number which is not an integer. This has the form $R'|R''\times\{0\}|\{\frac{1}{n}\} = \{\varepsilon r',\varepsilon r'' - \frac{1}{n}(r''-r)\}|\{\varepsilon r'',\varepsilon r' + \frac{1}{n}(r-r')\}$. By mutual cofinality this simplifies to $\varepsilon r',\varepsilon r''\}$. This enables us to obtain the pattern by induction. We obtain the sequence for ε followed by the tail of the sequence for r after the first $+$. For example $\frac{3}{4} = (+-+)$. Hence $\frac{3}{4}\varepsilon$ is the sequence for ε followed by $(-+)$. To justify the induction we need to verify the pattern for positive integers as well. In that case R'' is empty and the expression simplifies instead to $\{\varepsilon r'\}|\{\frac{1}{n}\}$. However, the pattern works in any case although here there are no infinitesimal upper elements as before.

A natural expression to consider next is $\varepsilon^2 = \frac{1}{\omega^2}$. In analogy with $\varepsilon = \frac{1}{\omega}$ one natural candidate for the sequence is a plus followed by ω^2 minuses. But $\varepsilon^2 = \varepsilon.\varepsilon = (\{0\}|\{\frac{1}{n}\})\times(\{0\}|\{\frac{1}{n}\}) = \{0,\frac{\varepsilon}{n} + \frac{\varepsilon}{n} - \frac{1}{n^2}\}|\{\frac{\varepsilon}{n}\} = \{0\}|\{\frac{\varepsilon}{2n}\}$. It follows from the previous result for rw that the result is the sequence consisting of a single plus followed by $\omega\cdot 2$ minuses. In particular, we do <u>not</u> have ω^2 minuses as one may naively guess. Of course, the contrast between the behaviour of $r\omega$ and $r\varepsilon$ should make an alert reader suspicious of such a guess at the outset.

It is next of interest to investigate $\varepsilon+\varepsilon^2$. Is it the sequence for ε followed by the sequence for ε^2? Well $\{0\}|\{\frac{1}{n}\} + \{0\}|\{\frac{\varepsilon}{m}\} = \{\varepsilon,\varepsilon^2\}|\{\varepsilon + \frac{\varepsilon}{m},\varepsilon^2 + \frac{1}{n}\} = \{\varepsilon\}|\{d\varepsilon\}$ where d is the set of all dyadic fractions larger than 1. Using the previous result for $r\varepsilon$ thus leads to the sequence for ε followed by the sequence for ε! Thus, again, juxtaposition behaves subtly. The ε^2 term is contributed by the annexation of the term for ε.

The examples we have done illustrate the most basic tricky phenomena that occur in attempting to find the sign sequence for various algebraic expressions. We close by considering the expression $\sqrt{\omega}$.

We ignore our previous construction of square roots and use a good guess method. Let b be the sequence consisting of ω pluses followed by ω^2 minuses. We consider the canonical representation and obtain a cofinal upper set by restricting ourselves to those sequences for which the number of minuses is a multiple of ω. So by earlier computations we have $b = \{n\}|\{\frac{\omega}{2m}\} = \{n\}\}\{\frac{\omega}{m}\}$. Note first that this is <u>a priori</u> a plausible candidate for $\sqrt{\omega}$ since $n < \sqrt{\omega}$ and $\sqrt{\omega} = \frac{\omega}{\sqrt{\omega}} < \frac{\omega}{n}$.

$$b^2 = \{bn_1 + bn_2 - n_1n_2, \frac{b\omega}{m_1} + \frac{b\omega}{m_2} - \frac{\omega^2}{m_1m_2}\}|\{nb + \frac{\omega b}{m} - \frac{\omega n}{m}\}.$$

Now $\omega = \{n\}|\phi$. So we must verify the conditions for the cofinality theorem. As usual in a proof of this form all we know about b in advance is that $n < b < \frac{\omega}{n}$, i.e. that b and $\frac{\omega}{b}$ are both infinite. There is a trap which is a temptation to use circular reasoning (i.e., that $b^2 = \omega$). In this respect this proof is similar to the one dealing with $\frac{1}{\omega}$ and the second proof for the sign sequence for $\frac{1}{2}\omega$.

The conditions are easily verified.

$bn_1 + bn_2 - n_1n_2 \leq b(n_1+n_2) < \omega$ since $\frac{\omega}{b}$ is infinite. For the same reason $\frac{b\omega}{m_1} + \frac{b\omega}{m_2} - \frac{\omega^2}{m_1m_2} < 0 < \omega$. Also $nb + \frac{\omega b}{2m} - \frac{\omega n}{2m} \geq \frac{\omega(b-n)}{2m} > \omega$ since b is infinite.

This verifies the betweenness condition. If we let $n_1 = n_2 = 1$, we obtain $2b-1$ as a lower element which clinches the cofinality condition since b is infinite. In fact, the same element $2b-1$ works for all n. This shows that $b^2 = \omega$.

The supply of surreal numbers is very rich. Continuing in the above manner is like using a teaspoon to empty an ocean. It is time now to get some sort of hold on more general elements.

5 NORMAL FORM

A COMBINATORIAL LEMMA ON SEMIGROUPS

So far we have more or less accepted everything we needed from outside the theory of surreal numbers since the material was very elementary. However, in order to study the normal form we need a combinatorial lemma which is not as well known, and since it is interesting in its own right we shall prove it here.

Lemma 5.1. Let T be a set of positive well-ordered elements in a linearly-ordered semigroup. Then the set S of finite sums of elements of T is also well-ordered and each element of S can be expressed as a sum of elements of T in only a finite number of ways.

Proof. Both parts will follow if we show that any sequence $\{s_n\}$ of elements of S such that $s_n \geq s_{n+1}$ for all n in which the representations are distinct must eventually terminate. We will assume an infinite sequence and obtain a contradiction.

Case 1. Suppose that we have only binary sums. Let $s_n = a_n+b_n$ where $a_n \in T$ and $b_n \in T$. Since T is well ordered, there exists a subsequence a_{i_n} of a_n such that $a_{i_{n+1}} \geq a_{i_n}$ for all n. We can obtain such a subsequence as follows. Choose i such that a_i is the least value of the a's. If $i_1 < i_2 \ldots < i_n$ have been chosen, then choose $i_{n+1} > i_n$ such that $a_{i_{n+1}}$ is the least value of the a's with index larger than i_n. Similarly there exists a subsequence of b_{j_n} of b_{i_n} such that $b_{j_{n+1}} \geq b_{j_n}$ for all n. Then $a_{j_{n+1}} \geq a_{j_n}$ and $b_{j_{n+1}} \geq b_{j_n}$ for all n. Since $a_{j_{n+1}} + b_{j_{n+1}} \leq a_{j_n} + b_{j_n}$, we obtain that

$a_{j_{n+1}} = a_{j_n}$ and $b_{j_{n+1}} = b_{j_n}$, contradicting the hypothesis that the representations are distinct.

Case 2. Suppose that we have only sums of k terms for fixed k. Then the proof is essentially the same, but the notation is minutely more complicated. If s_n has the form $a_{n1} + a_{n2} \ldots a_{nk}$ then we simply iterate the process of taking subsequences for all fixed $j \le k$.

Case 3. If only a finite number of k's occur, then we can reduce to case 2 since at least one of these k's must appear infinitely often.

Case 4. Suppose an infinite number of k's occur, i.e. the value of k- is unbounded. Then we can choose a subsequence $\{s_n'\}$ of $\{s_n\}$ as follows: $s_i' = a_{i1} + a_{i2} \ldots + a_{in_i}$ where n_i is a strictly increasing function of i. In particular, $n_i \ge i$. Also, we express the sums in non-increasing order, i.e. $a_{i1} \ge a_{i2} \ge \ldots \ge a_{in_i}$. We now choose a subsequence of $\{s_n'\}$ such that a_{i1} is monotonic increasing, a subsequence such that a_{i2} is monotonic increasing, etc. By the usual diagonal method we obtain a subsequence s_n'' such that a_{ij} for fixed j is monotonic increasing as a function of i for $i \ge j$. [Since $n_i \ge i$, a_{ij} is defined for $i \ge j$.] First we show that necessarily $n_i > i$ for all i. Suppose $n_i = i$. Then consider

$$s_i'' = a_{i1} + a_{i2} \ldots a_{ii}$$

and

$$s_{i+1}'' = a_{i+1,1} + a_{i+1,2} \ldots a_{i+1,i} + a_{i+1,i+1} \ldots .$$

Since $a_{i+1,j} \ge a_{i,j}$ for all $j \le i$, and s_{i+1}'' contains an extra term $a_{i+1,i+1}$ which has no analogue in s_i'', it follows that $s_{i+1}'' > s_i''$. This contradicts the assumption that the sequence is monotonic decreasing.

For arbitrary i we now compare:

$$s_i'' = a_{i1} + a_{i2} \ldots a_{ii} + a_{i,i+1} \ldots a_{in_i}$$

with

$$s_{n_i}'' = a_{n_i1} + a_{n_i2} \ldots a_{n_ii} + a_{n_i,i+1} \ldots a_{n_in_i} + \ldots .$$

Since $a_{n_ij} \ge a_{ij}$ for $j \le i$, since s_n'' contains more terms than s_i'' and since s_n'' is monotonic decreasing, there must exist k such that

$i < k \leq n_i$ with $a_{ik} > a_{n_i k}$. Hence $a_{ii} \geq a_{ik} > a_{n_i k} \geq a_{n_i n_i}$. By induction we may now define a function $b(i)$ as follows: $b(1) = 1$ and $b(i+1) = n_{b(i)}$. Then $\{a_{b(i),b(i)}\}$ is an infinite strictly decreasing sequence of elements of T, which is our final contradiction. This completes the proof.

B THE ω MAP

Up to now we have considered real numbers, ordinals, and algebraic combinations of these. What we need now is a more tractable way of looking at a general surreal number. We begin by studying orders of magnitude, a concept which has meaning in any linearly ordered field containing the real numbers.

We define an equivalence relation on the positive surreal numbers.

Definition. $a \sim b$ iff ($\exists$ integer n)($na \geq b$ and $nb \geq a$). This is trivially an equivalence relation. The equivalence classes are called "orders of magnitude." Related to this is another definition.

Definition. $a \gg b$ iff ($\forall$ integers n) ($nb \leq a$). $a \ll b$ iff $b \gg a$. We say in words that a has a higher order of magnitude than b. Clearly we have $a \gg b$, $b \gg a$, or $a \sim b$.

We shall assume that the reader has no trouble in seeing the most obvious consequences of these relations, so that they will be freely used without a detailed explanation when needed.
$a \ll b \Rightarrow na \ll b$, $a \ll c$ and $b \ll c \Rightarrow a+b \ll c$. One property of special interest is $a \sim b$ and $a < c < b \Rightarrow a \sim c$, i.e. the equivalence classes are convex.

One basic fact which is special about the surreal numbers is that each equivalence class has a canonical number.

Theorem 5.1. Let a be a positive surreal number. Then there exists a unique x of minimal length such that $x \sim a$.

Proof. The argument uses only the convexity. Because of well-ordering there certainly exists an x of minimal length such that $x \sim a$.

Suppose x and y are distinct, both have minimal length and $x \sim a \sim y$. Let z be the common initial segment. By the convexity property $z \sim a$. Also $\ell(z) < \ell(x)$ which contradicts the minimality of $\ell(x)$.

Remark. The same argument shows that the element of minimal length is an initial segment of every other element equivalent to a.

Similarly to the above one can define additive orders of magnitude using addition rather than multiplication.

Notation. $x \sim a$ iff $(\exists$ integer $n)(a+n \geq b)$ and $(b+n \geq a)$.

Since this is less important for our purpose and, besides, is similar to the case of multiplicative orders of magnitude including the possession of the convexity property, we do not give the details. It suffices to note that here also every equivalence class has a canonical member which is defined in a similar way.

We now come to one of Conway's most remarkable discoveries [1, page 31]. The canonical elements can themselves be parametrized by the surreal numbers in a natural way.

For every surreal number b we shall define an element written ω^b which may be thought of as the canonical element of the bth order of magnitude. (Although there are philosophical objections to the use of the exponential notation which have some validity, there is enough in common with exponentiation to make the notation psychologically convenient.) As usual, we use induction and assume that ω^c has been defined for all proper segments of b. Then.

Definition. $\omega^b = \{0, r\omega^{b'}\}|\{s\omega^{b''}\}$ where r and s run through the set of all positive real numbers, and b' and b'' run as in our usual notation through the lower and upper elements of the canonical representation of b.

By confinality we can, of course, limit r to integers and s to dyadic fractions with numerator 1.

Theorem 5.2. ω^b is always defined and greater than 0. Furthermore, $b < c \to \omega^b << \omega^c$.

Proof. We prove this by induction on the length of b. Since $b' < b''$ we have $\omega^{b'} \ll \omega^{b''}$ by the inductive hypothesis. Hence, for all positive reals r and s, $r\omega^{b'} < s\omega^{b''}$. Also, $0 < s\omega^{b''}$. Hence ω^b is defined. Since 0 is a lower element in the definition, $0 < \omega^b$.

To conclude the proof, we use a method which is similar to the one we used for the arithmetical operations. Here the computation is immediate. Suppose $b < c$ and d is the common initial segment. If $d = b$ or c then the conclusion is immediate from the definition. Otherwise we have $\omega^b \ll \omega^d \ll \omega^c$.

Corollary. The uniformity theorem holds for ω^b, i.e. if $b = F|G$ for an arbitrary representation the same formula holds, i.e. $\omega^b = \{0, r\omega^F\}|\{s\omega^G\}$. ($\omega^F = \{\omega^x : x \in F\}$ and similarly for G.)

Proof. As usual, this follows from the inequality $b < c \rightarrow \omega^b < \omega^c$ using the inverse cofinality and the cofinality theorems.

Theorem 5.3. An element has the form ω^b if and only if it is the element of minimal length in an equivalence class under $\sim$.

Proof. First consider an element of the form ω^b. We have $\omega^b = \{0, r\omega^{b'}\}|\{s\omega^{b''}\}$. If $x \sim \omega^b$ then x also satisfies $r\omega^{b'} < x < s\omega^{b''}$ since r and s are arbitrary positive reals. Hence ω^b is an initial segment of x, so $\ell(\omega^b) \leq \ell(x)$.

For the converse, we show that every positive element is equivalent to an element of the form ω^b. In view of the inequality $b < c \rightarrow \omega^b \ll \omega^c$, such an element is unique if it exists. We use induction. Let $a = A'|A''$. Then $0 \in A'$. By the inductive hypothesis every element in $A' \cup A''$ is equivalent to an element of the form ω^b. Let $F = \{y : (\exists x \in A')(x \sim \omega^y)\}$ and $G = \{y : (\exists x \in A'')(x \sim \omega^y)\}$. Suppose $F \cap G \neq \emptyset$ and let $y \in F \cup G$. Then $\omega^y \sim x \in A'$ and $\omega^y \sim z \in A''$. Since $x < a < z$ it follows that $a \sim \omega^y$.

Now suppose $F \cap G = \phi$. We claim that $F < G$. For suppose $x \in F$, $y \in G$ and $x > y$. Then $\omega^x \sim a' \in A'$ and $\omega^y \sim a'' \in A''$. Hence $x > y \rightarrow \omega^x \gg \omega^y \rightarrow a' \gg a''$. This is impossible since $a' < a''$. Since $F < G$, $F|G$ has meaning. Let $z = F|G$. Then ω^F is a complete set of representatives for the equivalence classes containing the elements of $A' - \{0\}$ and similarly for ω^G with respect to A''. We now consider three cases.

Case 1. $r\omega^x \geq a$ for some positive real r and some $x \in F$. Let $a' \in A'$ satisfy $a' \sim \omega^x$. Then $a' \leq a \leq r\omega^x$ but $a' \sim \omega^x \sim r\omega^x$; hence $a \sim \omega^x$.

Case 2. $r\omega^x \leq a$ for some positive real r and some $x \in G$. Let $a'' \in A''$ satisfy $a'' \sim \omega^x$. Then $r\omega^x \leq a \leq a''$ but $r\omega^x \sim \omega^x \sim a''$; hence $a \sim \omega^x$.

Case 3. Neither case 1 nor case 2 is satisfied. This says that $r\omega^F < a < s\omega^G$. Now let $a' \in A' - \{0\}$. Then $(\exists x\varepsilon F)(a' \sim \omega^x)$. In particular, for some real r, $r\omega^x \geq a'$. Similarly for $a'' \in A''$ we have $(\exists x\varepsilon G)(a'' \sim \omega^x)$. Hence for some positive real s, $s\omega^x \leq a''$. Since $a = A'|A''$ this shows that the cofinality condition is satisfied for $\{0, r\omega^F\}|\{s\omega^G\}$. Hence $a = \{0, r\omega^F\}|\{s\omega^G\} = \omega^z$.

The theorem now follows immediately, for if a has minimal length in its equivalence class, then $a \sim \omega^b \to a = \omega^b$ since, as we have already shown, ω^b has the minimal length property.

Our next result gives some justification for the exponential notation.

Theorem 5.4. (a) $\omega^0 = 1$, (b) $\omega^a\omega^b = \omega^{a+b}$, (c) for ordinals a our ω^a is the same as the ordinal ω^a in the usual sense.

Proof. By definition $\omega^0 = \{0\}|\phi = 1$. We prove (b) by induction as usual. Let $a = A'|A''$ and $b = B'|B''$. Then by the formula for addition and the uniformity theorem, using the facts that $\omega^a = \{0, r\omega^{a'}\}|\{s\omega^{a''}\}$ and $\omega^b = \{0, r_1\omega^{b'}\}|\{s_1\omega^{b''}\}$, we obtain that $\omega^{a+b} = \{0, r\omega^{a'+b}, r_1\omega^{a+b'}\}|\{s\omega^{a''+b}, s_1\omega^{a+b''}\}$. Similarly for multiplication we obtain that

$$\omega^a\omega^b = \{0, r\omega^{a'}\omega^b, r_1\omega^{b'}\omega^a, r\omega^{a'}\omega^b + r_1\omega^{b'}\omega^a - rr_1\omega^{a'}\omega^{b'}, s\omega^{a''}\omega^b + s_1\omega^{b''}\omega^a - ss_1\omega^{a''}\omega^{b''}\}|\{s\omega^{a''}\omega^b, s_1\omega^{b''}\omega^a, r\omega^{a'}\omega^b + s_1\omega^{b''}\omega^a - rs_1\omega^{a'}\omega^{b''}, s\omega^{a''}\omega^b + r_1\omega^{b'}\omega^a - r_1s\omega^{b'}\omega^{a''}\}.$$

By the inductive hypothesis this may be written

$$\{0, r\omega^{a'+b}, r_1\omega^{b'+a}, r\omega^{a'+b} + r_1\omega^{b'+a} - rr_1\omega^{a'+b'}, s\omega^{a''+b} + s_1w^{b''+a} - ss_1\omega^{a''+b''}\}|\{s\omega^{a''+b}, s_1\omega^{b''+a}, r\omega^{a'+b} + s_1\omega^{b''+a} - rs_1\omega^{a'+b''}, s\omega^{a''+b}$$

$+ r_1\omega^{b'+a} - r_1 s\omega^{b'+a''}\}$.

We now show that the representations for ω^{a+b} and $\omega^a\omega^b$ are mutually cofinal. One direction is immediate since the terms for ω^{a+b} are among the terms for $\omega^a\omega^b$. Since $b > c \rightarrow \omega^b >> \omega^c$ the other direction follows easily by elementary reasoning with orders of magnitude. First $r\omega^{a'+b} + r_1\omega^{b'+a} - rr_1\omega^{a'+b'} \leq (r+r_1)\omega^{\max(a'+b,a+b')}$. Next, $s\omega^{a''+b} + s_1\omega^{b''+a} - ss_1\omega^{a''+b''} < 0$ because the term containing $\omega^{a''+b''}$ dominates the other terms. Now we consider a typical upper element $r\omega^{a'+b} + s_1\omega^{b''+a} - rs_1\omega^{a'+b''}$. Since the term containing $\omega^{b''+a}$ dominates the other terms, it follows that for any $s_2 < s_1$ the element is above $s_2\omega^{b''+a}$. Since the same argument applies if a and b are interchanged, this verifies the cofinality.

(c) follows easily by induction. Let $a = \{a'\}|\phi$. For the purpose of this proof let us temporarily use $F(c)$ instead of ω^c for our order of magnitude, and use ω^c in its usual sense in the theory of ordinals. Then $F(a) = \{0,F(a')\}|\phi = \{0,n\omega^{a'}\}|\phi = \omega^a$. The last equality is a basic fact concerning the ordering of the ordinals. This completes the proof.

<u>Corollary</u>. $\omega^a\omega^{-a} = 1$.

Theorem 5.4 gives some justification for the exponential notation. However, the main justification comes from the nice way the operations behave on the normal forms of the surreal numbers which we will see later.

<u>Remark</u>. Note that if F contains no maximum, then $r\omega^F$ may be replaced simply by ω^F because of cofinality. For suppose $r\omega^{x'} \in F$. If $y' \in F$ and $y' > x'$ then $\omega^{y'} >> \omega^{x'}$; hence $\omega^{y'} > r\omega^{x'}$. A similar remark applies to $s\omega^G$.

C NORMAL FORM

We now obtain something which is analogous to the normal form for ordinals. Here we need transfinite sums. We shall define expressions of the form $\sum_{i<\alpha} \omega^{a_i} r_i$ where (a_i) is a strictly decreasing transfinite sequence of order type α and r_i is a real number distinct from 0 for all i. This is done inductively on α.

Case 1. α is a non-limit ordinal. Let $\alpha = \beta+1$. Then
$\sum_{i<\alpha} \omega^{a_i} r_i = (\sum_{i<\beta} \omega^{a_i} r_i) + \omega^{a_\beta} r_\beta$.

Case 2. α is a limit ordinal. We obtain $\sum_{i<\alpha} \omega^{a_i} r_i$ in the form $F|G$.

A typical element of F has the form $\sum_{i\leq\beta} \omega^{a_i} s_i$, where $\beta < \alpha$ such that $s_i = r_i$ for $i < \beta$, and $s_\beta = r_\beta - \varepsilon$ where ε is positive real.

Similarly a typical element of G has the form $\sum_{i\leq\beta} \omega^{a_i} s_i$, where $\beta < \alpha$, such that $s_i = r_i$ for $i < \beta$, and for $s_\beta = r_\beta + \varepsilon$ where ε is positive real. (We use the natural notation $\sum_{i\leq\beta} \omega^{a_i} r_i$ as an alternative for $\sum_{i<\beta+1} \omega^{a_i} s_i$.)

If 0 is regarded as a limit ordinal the definition leads to the empty sum being $\phi|\phi = 0$. For α finite, the expression is just the ordinary finite sum. For α infinite, a proof that $F < G$ is needed in order to show that the definition makes sense. In fact, we shall show that the ordering on surreal numbers is consistent with the lexicographic ordering with respect to the a's and r's in the normal form. First we define the lexicographical order on expressions $\sum_{i<\alpha} \omega^{a_i} r_i$.

Let $x = \sum_{i<\alpha} \omega^{a_i} r_i$ and $y = \sum_{i<\beta} \omega^{b_i} s_i$.

Let γ be the least ordinal such that $(a_\gamma, r_\gamma) \neq (b_\gamma, s_\gamma)$. If $\gamma < \min(\alpha,\beta)$ then $x > y$ iff $a_\gamma > b_\gamma$ or $a_\gamma = b_\gamma$ and $r_\gamma > s_\gamma$. If $\gamma = \beta$, then $x > y$ iff $r_\gamma > 0$. If $\gamma = \alpha$ then $x > y$ iff $s_\gamma < 0$.

Note that this is consistent with the situation for the normal form for ordinals.

Theorem 5.5. The expression $\sum_{i<\alpha} \omega^{a_i} r_i$ is defined for all strictly decreasing sequences (a_i) and all nonzero real r_i. The ordering is given by the lexicographical order. Furthermore for all $\beta < \alpha$,

$\left|\sum_{i<\alpha} \omega^{a_i} r_i - \sum_{i<\beta} \omega^{a_i} r_i\right| \ll \omega^{a_j}$ if $j < \beta$. (We call the latter inequality the "tail property".)

<u>Proof</u>. We use induction on α.

<u>Case 1</u>. α is a non-limit ordinal. Let $\alpha = \beta+1$. Then $\sum_{i<\alpha} \omega^{a_i} r_i = \sum_{i<\beta} \omega^{a_i} r_i + \omega^{a_\beta} r_\beta$.

To begin with, let $x = \sum_{i<\alpha} \omega^{a_i} r_i$ and $y = \sum_{i<\alpha} \omega^{b_i} s_i$ and suppose $x > y$ in the lexicographical order. We must show that $x > y$ as surreal numbers. If $(\forall i<\beta)[(a_i,r_i) = (b_i,s_i)]$ then either $a_\beta > b_\beta$ or $a_\beta = b_\beta$ and $r_\beta > s_\beta$. In either case $\omega^{a_\beta} r_\beta > \omega^{b_\beta} s_\beta$ by elementary reasoning with orders of magnitude. Since addition preserves order, it follows that $x > y$.

Next assume $(a_\gamma,r_\gamma) \neq (b_\gamma,s_\gamma)$ for some $\gamma < \beta$ but $(a_\delta,r_\delta) = (b_\delta,s_\delta)$ for $\delta < \gamma$. Then either $a_\gamma > b_\gamma$ or $a_\gamma = b_\gamma$ and $r_\gamma > s_\gamma$. In either case $\omega^{a_\gamma} r_\gamma - \omega^{b_\gamma} s_\gamma \geq \omega^{a_\gamma} t$ for some positive real t. Hence $\sum_{i<\gamma+1} \omega^{a_i} r_i - \sum_{i<\gamma+1} \omega^{b_i} s_i = (\sum_{i<\gamma} \omega^{a_i} r_i + \omega^{a_\gamma} r_\gamma) - (\sum_{i<\gamma} \omega^{b_i} s_i + \omega^{b_\gamma} s_\gamma)$ $= \omega^{a_\gamma} r_\gamma - \omega^{b_\gamma} s_\gamma \geq \omega^{a_\gamma} t$. However $\sum_{i<\beta} \omega^{a_i} r_i$ and $\sum_{i<\beta} \omega^{b_i} s_i$ have the tail property by the inductive hypothesis. Hence

$\left|\sum_{i<\beta} \omega^{a_i} r_i - \sum_{i<\gamma+1} \omega^{a_i} r_i\right| \ll \omega^{a_\gamma}$ and $\left|\sum_{i<\beta} \omega^{b_i} s_i - \sum_{i<\gamma+1} \omega^{b_i} s_i\right| \ll \omega^{a_\gamma}$.

Therefore $\sum_{i<\beta} \omega^{a_i} r_i - \sum_{i<\beta} \omega^{b_i} s_i \geq \omega^{a_\gamma} t'$ for a positive real t'. (We can use any t' less than t.) Again, since $\omega^{b_\beta} \ll \omega^{b_\gamma} \leq \omega^{a_\gamma}$, we obtain $x - y = \sum_{i<\alpha} \omega^{a_i} r_i - \sum_{i<\alpha} \omega^{b_i} s_i = \sum_{i<\beta} \omega^{a_i} r_i - \sum_{i<\beta} \omega^{b_i} s_i + \omega^{a_\beta} r_\beta - \omega^{b_\beta} s_\beta \geq \omega^{a_\gamma} t''$ for a positive real t''. In particular $x > y$.

The same argument applies if only one of x and y have a representation of length α. A slight difference in notation is needed if one is picayune since a_β and b_β are not both present, but in any case the superiority which x gains in the γth stage is necessarily maintained since whichever one of the above is present is still of lower order of magnitude than ω^{a_γ}. (There is even a possible case where a_γ is not present. Then we use ω^{b_γ} instead.)

We now check the tail property. Let $\gamma < \alpha$ and $j < \gamma$. Consider $|\sum_{i<\alpha} \omega^{a_i} r_i - \sum_{i<\gamma} \omega^{a_i} r_i|$. This is $|\omega^{a_\beta} r_\beta + (\sum_{i<\beta} \omega^{a_i} r_i - \sum_{i<\gamma} \omega^{a_i} r_i)|$. If $\gamma = \beta$ this reduces to $|\omega^{a_\beta} r_\beta|$ which is certainly of lower magnitude than ω^{a_j}. Otherwise $|\omega^{a_\beta} r_\beta|$ is certainly still of lower magnitude than ω^{a_j}, and so is $|\sum_{i<\beta} \omega^{a_i} r_i - \sum_{i<\gamma} \omega^{a_i} r_i|$ by the inductive hypothesis. Hence the absolute value of the sum is also, and we are done.

<u>Case 2</u>. α is a limit ordinal.

By the inductive hypothesis the ordering for elements in $F \cup G$ is given by the lexicographical ordering; hence $F < G$.

We now verify the tail property. Let $\beta < \alpha$ and $j < \beta$. We must consider $|\sum_{i<\alpha} \omega^{a_i} r_i - \sum_{i<\beta} \omega^{a_i} r_i|$. Now among the typical elements of F in the representation of $\sum_{i<\alpha} \omega^{a_i} r_i$ is $\sum_{i<j} \omega^{a_i} r_i - \omega^{a_j} \varepsilon$. This is immediate from the definition, since $\sum_{i<j} \omega^{a_i} r_i + \omega^{a_j} r_j = \sum_{i\leq j} \omega^{a_i} r_i$. (One must be cautious in reasoning with these infinite "sums". Other results which may appear to be just as obvious might require a technical proof because of our specialized definition of infinite sums.) Similarly, among the typical elements of G is $\sum_{i<j} w^{a_i} r_i + \omega^{a_j} \varepsilon$. Hence $\sum_{i<j} \omega^{a_i} r_i - \omega^{a_j} \varepsilon < \sum_{i<\alpha} \omega^{a_i} r_j < \sum_{i<j} \omega^{a_i} r_i + \omega^{a_j} \varepsilon$. By the lexicographical order, and the inductive hypothesis, we have $\sum_{i<j} \omega^{a_i} r_i - \omega^{a_j} \varepsilon < \sum_{i<\beta} \omega^{a_i} r_i < \sum_{i<j} \omega^{a_i} r_i + \omega^{a_j} \varepsilon$. Therefore $|\sum_{i<\alpha} \omega^{a_i} r_i - \sum_{i<\beta} \omega^{a_i} r_i| < \omega^{a_j}(2\varepsilon)$. Since ε is an arbitrary positive real this shows that $\sum_{i<\alpha} \omega^{a_i} r_i - \sum_{i<\beta} \omega^{a_i} r_i| << \omega^{a_j}$ as desired. The proof that the lexicographical order is the correct order is similar to the proof in the non-limit ordinal case. As before, let $x = \sum_{i<\alpha} \omega^{a_i} r_i$

and $y = \sum_{i<\alpha} \omega^{b_i} s_i$ and suppose $x > y$ in the lexicographical order. Suppose $(a_\delta, r_\delta) = (b_\delta, s_\delta)$ for all $\delta < \gamma$ for some $\gamma < \alpha$ but $(a_\gamma, r_\gamma) \neq (b_\gamma, s_\gamma)$. Then $\sum_{i\leq\gamma} \omega^{a_i} r_i - \sum_{i\leq\gamma} \omega^{b_i} s_i \geq \omega^{a_\gamma} t$ for some positive real t. By the tail property,
$|\sum_{i<\alpha} \omega^{a_i} r_i - \sum_{i\leq\gamma} \omega^{a_i} r_i| \ll \omega^{a_\gamma}$ and $|\sum_{i<\alpha} \omega^{b_i} s_i - \sum_{i\leq\gamma} \omega^{b_i} s_i| \ll \omega^{b_\gamma} \leq \omega^{a_\gamma}$.
Therefore $\sum_{i<\alpha} \omega^{a_i} r_i - \sum_{i<\alpha} \omega^{b_i} s_i \geq \omega^{a_\gamma} t'$ for some positive real t'.

If only one of x and y has a representation of length α, then the earlier remark in the non-limit ordinal case remains valid. This completes the proof.

The next theorem gives us the importance of the transfinite sums we have been discussing.

Theorem 5.6. Every surreal number can be expressed uniquely in the form $\sum_{i<\alpha} \omega^{a_i} r_i$.

Proof. Uniqueness is immediate from the fact that the ordering is given by the lexicographical ordering.

Now let x be an arbitrary non-zero surreal number. We know by Theorem 5.3 that $|x| \sim \omega^a$ for some a. Let S be the set of all real numbers s such that $s\omega^a \leq x$. Since $|x| \sim \omega^a$, S is non-empty and is bounded above. Let $r = \ell.\text{u.b. } S$. Then $(r+\varepsilon)\omega^a > x$ and $(r-\varepsilon)\omega^a < x$ for all positive real ε; hence $|x-\omega^a r| \ll \omega^a$. Since $|x| \sim \omega^a$ it follows that $r \neq 0$. It is clear that the above property determines r uniquely. For convenience in the proof we shall use the notation $A(x) = \omega^a r$ if $x \neq 0$.

Now assume that x cannot be expressed in the form $\sum_{i<\alpha} \omega^{a_i} r_i$. We define a sequence (a_i, r_i) where i runs through all the ordinals. Suppose (a_i, r_i) is defined for all $i < \alpha$. Then $A[x - \sum_{i<\alpha} \omega^{a_i} r_i] = \omega^{a_\alpha} r_\alpha$.

Intuitively speaking, we are getting better and better approximations to x as α is increasing. We first show that the a's are decreasing, so that the sums make sense.

First let $\alpha = \beta+1$. Then $A[x-\sum_{i<\beta}\omega^{a_i}r_i] = \omega^{a_\beta}r_\beta$.

$\omega^{a_\alpha}r_\alpha = A[x-\sum_{i<\alpha}\omega^{a_i}r_i] = A[(x-\sum_{i<\beta}\omega^{a_i}r_i) - \omega^{a_\beta}r_\beta] \ll \omega^{a_\beta}$ by the inductive definition of (a_β,r_β). Hence $a_\alpha < a_\beta$.

Now let α be a limit ordinal and let $\beta < \alpha$. We already know that $|x-\sum_{i\leq\beta}\omega^{a_i}r_i| \sim A[x-\sum_{i\leq\beta}\omega^{a_i}r_i] \ll \omega^{a_\beta}$. By the tail property $|\sum_{i<\alpha}\omega^{a_i}r_i - \sum_{i\leq\beta}\omega^{a_i}r_i| \ll \omega^{a_\beta}$. Hence $|x-\sum_{i<\alpha}\omega^{a_i}r_i| \ll \omega^{a_\beta}$. Therefore $\omega^{a_\alpha}r_\alpha = A(x-\sum_{i<\alpha}\omega^{a_i}r_i) \ll \omega^{a_\beta}$. So finally $a_\alpha < a_\beta$.

Since by hypothesis x cannot be expressed as a sum, $\sum_{i<\alpha}\omega^{a_i}r_i$ has meaning for all α.

We next show that $\ell(\sum_{i<\alpha}\omega^{a_i}r_i) \geq \alpha$ for any general sum. Although this inequality is crude it suffices for our immediate purpose.

Let $\alpha < \beta$. Then the elements of F and G used in the representation of $\sum_{i<\alpha}\omega^{a_i}r_i$ are also used in the representation of $\sum_{i<\beta}\omega^{a_i}r_i$. Hence the former is an initial segment of the latter and thus it has smaller length. The uniqueness of representations as sums guarantees that the length is strictly smaller. This is enough to verify the inequality since every strictly increasing function f from ordinals to ordinals necessarily satisfies $f(x) \geq x$ for all x.

If α is a limit ordinal we already know that $|x - \sum_{i<\alpha}\omega^{a_i}r_i| \ll \omega^{a_\beta}$ for any $\beta < \alpha$. This shows that x satisfies $F < x < G$ for the F and G used in the representation of $\sum_{i<\alpha}\omega^{a_i}r_i$. Hence $\ell(\sum_{i<\alpha}\omega^{a_i}r_i) < \ell(x)$. This is true for all α. In view of the earlier inequality this implies that $\ell(x)$ is above every ordinal, which is absurd. This contradiction completes the proof.

We have now established the normal form for surreal numbers. The usual representation of ordinals in terms of powers of ω is a special case of this, since finite sums correspond to ordinary addition and this agrees with ordinal addition if terms are arranged so that there is no absorption.

Next we shall show the fundamental fact that the basic operations can be performed on elements in the normal form analogously to usual operations on polynomials. This is the main justification for the summation as well as the exponential notation.

If $x = \sum_{i<\alpha} \omega^{a_i} r_i$, we shall call α the normal length of x, abbreviated $n\ell(x)$. This is quite different from $\ell(x)$ which was defined at the beginning. We shall study $n\ell(x)$ in more detail later.

We now need some lemmas which will help us deal with the normal form.

<u>Lemma 5.2.</u> $\omega^a r = \{\omega^a(r-\varepsilon)\}|\{\omega^a(r+\varepsilon)\}$ where ε is an arbitrary positive real number.

We know that $r = \{r-\varepsilon\}|\{r+\varepsilon\}$ for any real r. Also, $\omega^a = \{0, s\omega^{a'}\}|\{t\omega^{a''}\}$ by definition where s and t are arbitrary positive real. Hence $\omega^a r$ has the form

$\{\omega^a(r-\varepsilon), \omega^a(r-\varepsilon)+(s\omega^{a'})\varepsilon, \omega^a(r+\varepsilon)-(t\omega^{a''})\varepsilon\}|$
$\{\omega^a(r+\varepsilon), \omega^a(r+\varepsilon)-(s\omega^{a'})\varepsilon, \omega^a(r-\varepsilon)+(t\omega^{a''}\varepsilon\}.$

Let ε_1 be positive, real, and less than ε. Since $\omega^{a'} \ll \omega^a \ll \omega^{a''}$, it is immediate that the lower terms are below $\omega^a(r-\varepsilon_1)$ and the upper terms above $\omega^a(r+\varepsilon_1)$. Hence the result follows by the cofinality theorem.

The above proof is a good typical example of reasoning with orders of magnitude and cofinality. This technique helps to give the normal form its tractability.

<u>Lemma 5.3.</u> $\sum_{i<\alpha} \omega^{a_i} r_i = \{\sum_{i<\alpha} \omega^{a_i} r_i + \omega^{a_\alpha}(r_\alpha-\varepsilon)\}|\{\sum_{i<\alpha} \omega^{a_i} r_i + \omega^{a_\alpha}(r_\alpha+\varepsilon)\}.$

<u>Proof.</u> Assume first that α is a limit ordinal. Then since $\sum_{i<\alpha} \omega^{a_i} r_i = \sum_{i<\alpha} \omega^{a_i} r_i + \omega^{a_\alpha} r_\alpha$ we can obtain a representation of $\sum_{i<\alpha} \omega^{a_i} r_i$ by using the definition for the first addend and lemma 5.2 for the second addend. Typical lower terms are $\sum_{i<\beta} \omega^{a_i} r_i - \omega^{\beta}\varepsilon + \omega^{a_\alpha} r_\alpha$ where $\beta < \alpha$ and $\sum_{i<\alpha} \omega^{a_i} r_i + \omega^{a_\alpha}(r_\alpha-\varepsilon)$. The latter terms are clearly cofinal by the

lexicographical order. Since the argument is similar for the upper terms terms, the result follows in this case.

In general, every ordinal α has the form $\alpha'+n$ where α' is a limit ordinal. We can now use induction on n. For convenience of notation we can assume the result for $\sum_{i<\alpha} \omega^{a_i} r_i$ and then prove it for $\sum_{i<\alpha+1} \omega^{a_i} r_i$. The proof is now similar to the above. The only difference is that the typical lower terms which are discarded because of cofinality now have the form $\sum_{i<\alpha} \omega^{a_i} r_i - \omega^{a_{\alpha}}\varepsilon$ by the inductive hypothesis; the upper terms are similar.

It is convenient to extend the definition of $\sum_{i<\alpha} \omega^{a_i} r_i$ to the case where r_i may take the value 0. In fact, we use exactly the same definition, but we of course no longer have unique representation.

<u>Lemma 5.4</u>. Let r_i be a sequence of length α, and let $\{n_i\}$ for $i < \beta$ be the subsequence of i's such that $r_i \neq 0$. Furthermore, suppose $b_i = a_{n_i}$ and $s_i = r_{n_i}$. Then $\sum_{i<\alpha} \omega^{a_i} r_i = \sum_{i<\beta} \omega^{b_i} s_i$.

<u>Proof</u>. Although this appears to be completely trivial, a proof is needed because of the special definition of infinite summation. Essentially we must show that including terms with $r_i = 0$ does not affect sums. We do this inductively on the length of the partial sums of $\sum_{i<\alpha} \omega^{a_i} r_i$. For a non-limit ordinal this is clear since we are dealing with ordinary addition. For limit ordinals caution is required since in the expression $\sum_{i<\alpha} \omega^{a_i} r_i$, the βth term plays a role even if $r_\beta = 0$, since it leads to elements in $F \cup G$ such as $\sum_{i<\beta} \omega^{a_i} r_i + \varepsilon\omega^{a_\beta}$. However, by the cofinality theorem we still obtain the same element. The trickiest case which occurs is the following: $\alpha = \beta+\omega$, $r_\beta \neq 0$ and $r_{\beta+n} = 0$ for all integers n. We want to show that $\sum_{i<\beta+\omega} \omega^{a_i} r_i = \sum_{i\leq\beta} \omega^{a_i} r_i$. The left-hand side is defined in terms of an F and G whereas the right-hand side is an ordinary sum. A typical lower term in the definition of $\sum_{i<\beta+\omega} \omega^{a_i} r_i$

is $\sum_{i<\gamma} \omega^{a_i} r_i - \omega^{\gamma}\varepsilon$ with $\gamma < \beta+\omega$, and there is a similar expression for a typical upper term. By lemma 5.3 the right-hand side is $\{\sum_{i<\beta} \omega^{a_i} r_i - \omega^{a_\beta}\varepsilon\}|\{\sum_{i<\beta} \omega^{a_i} r_i + \omega^{a_\beta}\varepsilon\}$. Hence by the cofinality theorem the right-hand side equals the left-hand side.

The case just considered represents a transition where we invoked lemma 5.3. In all other cases cofinality is all that is needed.

If we are given two surreal numbers the above lemma permits us to write them in the form $\sum_{i<\alpha} \omega^{a_i} r_i$ and $\sum_{i<\alpha} \omega^{a_i} s_i$, using the same α and a_i's, by inserting zeros where needed.

Lemma 5.5 (the associative law). $\sum_{i<\alpha+\beta} \omega^{a_i} r_i = \sum_{i<\alpha} \omega^{a_i} r_i + \sum_{j<\beta} \omega^{r_{\alpha+j}} r_{\alpha+j}$.

Proof. We use induction on β. If β is a non-limit ordinal this is just an instance of the ordinary associative law. If β is a limit ordinal we compute the right-hand side using the representations in the definition for the right addend, whereas for the left addend we use the representation in the definition or in Lemma 5.3 depending on whether α is a limit or non-limit ordinal. If $\sum_{j<\beta} \omega^{a_{\alpha+j}} r_{\alpha+j} = F|G$ then by cofinality the right-hand side may be expressed as $\{\sum_{i<\alpha} \omega^{a_i} r_i+F\}|\{\sum_{i<\alpha} \omega^{a_i} r_i+G\}$. (We invoke the lexicographical order and reason as in the proof of lemma 5.3. Thus a typical lower term has the form $\sum_{i<\alpha} \omega^{a_i} r_i + (\sum_{j<\gamma} \omega^{a_{\alpha+j}} r_{\alpha+j} - \varepsilon\omega^{a_{\alpha+\gamma}})$ for $\gamma < \beta$ which by the inductive hypothesis is $\sum_{i<\alpha+\gamma} \omega^{a_i} r_i - \varepsilon\omega^{a_{\alpha+\gamma}}$. A typical upper term can be written similarly. But this is cofinal in the representation for $\sum_{i<\alpha+\beta} \omega^{a_i} r_i$ so the proof is completed.

The above lemmas show that in spite of the apparently artificial definition of infinite sums they behave in many ways in a manner expected of sums. We now come to the important fact that formal polynomial addition works.

Theorem 5.7. $\sum_{i<\alpha} \omega^{a_i} r_i + \sum_{i<\alpha} \omega^{a_i} s_i = \sum_{i<\alpha} \omega^{a_i}(r_i+s_i)$.

Proof. First note that lemma 5.4 allows us to express the fact that formal polynomial addition works in this convenient form. As usual we use induction on α. If $\alpha = \beta+1$ this is immediate. In fact, $\sum_{i<\alpha} \omega^{a_i} r_i + \sum_{i<\alpha} \omega^{a_i} s_i = \sum_{i<\beta} \omega^{a_i} r_i + \omega^{a_\beta} r_\beta + \sum_{i<\beta} \omega^{a_i} s_i + \omega^{a_\beta} s_\beta$. By the inductive hypothesis this is $\sum_{i<\beta} \omega^{a_i}(r_i+s_i) + \omega^{a_\beta} r_\beta + \omega^{a_\beta} s_\beta = \sum_{i<\alpha} \omega^{a_i}(r_i+s_i)$ using the ordinary distributive law and the definition of summation.

Now suppose α is a limit ordinal. One typical lower element of the sum is $(\sum_{i\leq\beta} \omega^{a_i} r_i - \omega^{a_\beta}\varepsilon) + (\sum_{i<\alpha} \omega^{a_i} s_i)$ $= (\sum_{i\leq\beta} \omega^{a_i} r_i - \omega^{a_\beta}\varepsilon) + (\sum_{i\leq\beta} \omega^{a_i} s_i + \sum_{\beta<i<\alpha} \omega^{a_i} s_i)$ by lemma 5.5. ($\sum_{\beta<i<\alpha} \omega^{a_i} s_i$ has the natural meaning $\sum_{i<\alpha-(\beta+1)} \omega^{a_{\beta+1+i}} s_{\beta+1+i}$.) By the inductive hypothesis this is $\sum_{i\leq\beta} \omega^{a_i}(r_i+s_i) - (\omega^{a_\beta}\varepsilon) + \sum_{\beta<i<\alpha} \omega^{a_i} s_i$. By the lexicographical order this is mutually cofinal with $\sum_{i\leq\beta} \omega^{a_i}(r_i+s_i) - \omega^{a_\beta}\varepsilon$. By symmetry we obtain the same expression if we begin with a lower sum of the form $\sum_{i<\alpha} \omega^{a_i} r_i + (\sum_{i\leq\beta} \omega^{a_i} s_i - \omega^{a_\beta}\varepsilon)$, and we obtain a similar expression for a typical upper sum. But this gives us exactly $\sum_{i<\alpha} \omega^{a_i}(r_i+s_i)$ by definition, or by lemma 5.4, if $r_i+s_i = 0$ for some i.

We now turn to multiplication, and prove the remarkable fact that formal polynomial multiplication works. This is, of course, the main justification for the exponential notation and the normal form.

First we prove a special case which can be thought of as the infinite distributive law.

Lemma 5.6. $\omega^b [\sum_{i<\alpha} \omega^{a_i} r_i] = \sum_{i<\alpha} \omega^{b+a_i} r_i$.

Proof. We use induction on α. If $\alpha = \beta+1$ then we have $\omega^b [\sum_{i<\alpha} \omega^{a_i} r_i] = \omega^b [\sum_{i<\beta} \omega^{a_i} r_i + \omega^{a_\beta} r_\beta] = \omega^b [\sum_{i<\beta} \omega^{\beta_i} r_i] + \omega^b \omega^{a_\beta} r_\beta$ by the ordinary distributive law. By the inductive hypothesis and theorem 5.4 this is $\sum_{i<\alpha} \omega^{b+a_i} r_i + \omega^{b+a_\beta} r_\beta = \sum_{i<\alpha} \omega^{b+a_i} r_i$. Of course we need the fact that addition preserves order so that $(b+a_i)$ is also a strictly decreasing sequence for the above to make sense.

Now suppose α is a limit ordinal. We then compute the product using the standard representations $\omega^b = \{0, \omega^{b'} s\} | \{\omega^{b''} t\}$ and $\sum_{i<\alpha} \omega^{a_i} r_i = \{\sum_{i<\beta} \omega^{a_i} r_i - \omega^{a_\beta}\varepsilon\} | \{\sum_{i<\beta} \omega^{a_i} r_i + \omega^{a_\beta}\varepsilon\}$. In order to simplify the notation, if the latter is written in the form $d = \{d'\}|\{d''\}$ then both $d''-d$ and $d-d'$ have the form $\omega^{a_\beta}\varepsilon + c$ where $|c| \ll \omega^{a_\beta}$. Hence for $\varepsilon_1 < \varepsilon < \varepsilon_2$ they are between $\omega^{a_\beta}\varepsilon_1$ and $\omega^{a_\beta}\varepsilon_2$. We can write the product in the form $\{\omega^b d', \omega^b d' + \omega^{b'} s(d-d'), \omega^b d'' - \omega^{b''} t(d''-d)\} |$ $\{\omega^b d'', \omega^b d' + \omega^{b''} t(d-d'), \omega^b d'' - \omega^{b'} s(d''-d)\}$.

Since $\omega^{b''} \gg \omega^b \gg w^{b'}$, we obtain by elementary reasoning with orders of magnitude that

$\omega^{b''} t(d''-d) + \omega^{b'} s(d-d') \geq \omega^{b''} t\omega^{a_\beta}\varepsilon_1 > \omega^b[\omega^{a_\beta}(2\varepsilon_2)] \geq \omega^b[d''-d']$. Hence $\omega^b d'' - \omega^{b''} t(d''-d) \leq \omega^b d' + \omega^{b'} s(d-d')$. Similarly

$\omega^{b''} t(d-d') + \omega^{b'} s(d''-d) \geq \omega^b[d''-d']$; hence

$\omega^b d'' - \omega^{b'} s(d''-d) \leq \omega^b d' + \omega^{b''} t(d-d')$. Therefore by cofinality $\omega^b d$ can be expressed in the form

$\{\omega^b d', \omega^b d' + \omega^{b'} s(d-d')\} | \{\omega^b d'', \omega^b d'' - \omega^{b'} s(d''-d)\}$. Let d_1' and d_1'' be lower and upper elements respectively corresponding to ε_1 for the same β. Then $\omega^b(d_1' - d') = \omega^b[\omega^{a_\beta}(\varepsilon - \varepsilon_1)] > \omega^{b'} s\omega^{a_\beta}(\varepsilon_2) \geq \omega^{b'} s^y (d-d')$. Hence $\omega^b d_1' > \omega^b d' + \omega^{b'} s(d-d')$. Similarly $\omega^b(d' - d_1'') > \omega^{b'} s(d''-d)$. Hence $\omega^b d_1'' \leq \omega^b d' - \omega^{b'} s(d''-d)$.

Thus again we can simplify by cofinality and obtain $\omega^b d = \{\omega^b d'\} | \{\omega^b d''\}$.

Finally we can use the inductive hypothesis and obtain that $\omega^b d = \{\omega^b [\sum_{i<\beta} \omega^{a_i} r_i - \omega^{a_\beta}\varepsilon]\} | \{\omega^b [\sum_{i<\beta} \omega^{a_i} r_i + \omega^{a_\beta}\varepsilon]\} = \{\sum_{i<\beta} \omega^{b+a_i} r_i - \omega^{b+a_\beta}\varepsilon\} |$ $\{\sum_{i<\beta} \omega^{b+a_i} r_i + \omega^{b+a_\beta}\varepsilon\}$, which by definition is $\sum_{i<\alpha} \omega^{b+a_i} r_i$.

We are now ready to consider formal polynomial multiplication. If $x = \sum_{i<\alpha} \omega^{a_i} r_i$ and $y = \sum_{i<\beta} \omega^{b_i} s_i$ then we define the formal product $x \circ y$ to be $\sum_{\substack{i<\alpha \\ j<\beta}} \omega^{a_i+b_j} r_i s_j$. By lemma 5.1 each exponent a_i+b_j occurs only finitely many times and the set of all a_i+b_j is well-ordered. Hence the expression has meaning. (To be technical, when we consider an expression such as $\sum_{i<\alpha} \omega^{a_i} r_i$ we are applying the lemma to the positive elements a_0-a_i. This is adequate because we are dealing with binary products only. In more general situations we want a_0 to be 0 to avoid trouble.)

It is well-known and easy to verify that with respect to formal polynomial multiplication and addition one gets a ring. In fact, it is an ordered ring with respect to the ordering we have.

<u>Theorem 5.8</u> $(\sum_{i<\alpha} \omega^{a_i} r_i)(\sum_{i<\beta} \omega^{b_i} s_i) = \sum_{\substack{i<\alpha \\ j<\beta}} \omega^{a_i+b_j} r_i s_j$, i.e. the product agrees with the formal product.

<u>Proof</u>. Again we use induction. Also we tentatively use the symbol $\times$ for formal multiplication.

First suppose that either α or β is a non-limit ordinal. Assume $\alpha = \gamma+1$. The same argument applies if $\beta = \gamma+1$. Then

$(\sum_{i<\alpha} w^{a_i} r_i)(\sum_{i<\beta} \omega^{b_i} s_i) = (\sum_{i<\gamma} \omega^{a_i} r_i + \omega^{a_\gamma} r_\gamma)(\sum_{i<\beta} \omega^{b_i} s_i) = (\sum_{i<\gamma} \omega^{a_i} r_i)(\sum_{i<\beta} \omega^{b_i} s_i)$
$+ \omega^{a_\gamma} r_\gamma (\sum_{i<\beta} \omega^{b_i} s_i)$ by the ordinary distributive law. We now apply the inductive hypothesis to the left addend and lemma 5.6 to the right addend to obtain

$(\sum_{\substack{i<\gamma \\ j<\beta}} \omega^{a_i+b_j} r_i s_j + \sum_{i<\beta} \omega^{a_\gamma+b_i} r_\gamma s_i)$. The argument is now completed by theorem 5.7 which tells us that formal addition works.

Now suppose that α and β are both limit ordinals. We simplify the notation as follows: $y = \sum_{i<\alpha} \omega^{a_i} r_i$, $z = \sum_{i<\beta} \omega^{b_i} s_i$ and y^o, z^o are lower or upper elements in the representation of y and z

respectively in the basic definition. Then a typical lower or upper element of yz has the form $yz^o + y^oz - y^oz^o$. (As at times in the past we use a unified notation since the four kinds of terms involved are dealt with similarly.) By the inductive hypothesis this may be written $y\times z^o + y^o\times z - y^o\times z^o = (y\times z) - (y-y^o)\times(z-z^o)$ by ordinary algebra. (Recall that we already know that surreal addition agrees with formal addition.) Recall now that we get a lower element for yz if and only if y^o and z^o are on the same side.

Now $y-y^o$ has the form $\pm\omega^{a_\gamma}\varepsilon_1 + c_1$ for some $\gamma < \alpha$, some positive real ε_1 and some $|c_1| \ll \omega^{a_\gamma}$. Similarly $z-z^o$ has the form $\pm\omega^{a_\delta}\varepsilon_2 + c_2$ for some $\delta < \beta$, some positive real ε_2 and some $|c_2| \ll \omega^{a_\delta}$. Therefore $(y-y^o)\times(z-z^o)$ has the form $\pm\omega^{a_\gamma+a_\delta}\varepsilon_1\varepsilon_2 + c_3$, where $|c_3| \ll \omega^{a_\gamma+a_\delta}$. By the sign rule for multiplication for lower elements we have a plus in front of $\omega^{a_\gamma+a_\delta}$ and for upper elements a minus. By mutual cofinality we can now write $yz = (y\times z-\omega^{a_\gamma+a_\delta}\varepsilon)|(y\times z+\omega^{a_\gamma+a_\delta}\varepsilon)$. (Mutual cofinality follows from the observation that if $|c_1|, |c_2| \ll \omega^a$ and $r_1 < r_2$ are two real numbers then $\omega^a r_1 + c_1 < \omega^a r_2 + c_2$.)

We must now show that the right-hand side is $y\times z$. Now $y\times z = (y\times z-\omega^{a_\mu}\varepsilon)|(y\times z+\omega^{a_\mu}\varepsilon)$, where a_μ is a typical exponent in the series for $y\times z$. This again follows by cofinality. By lemma 5.3 this is valid even if $y\times z$ has a last term. Now every exponent in $y\times z$ has the form $a_\gamma+a_\delta$ (though the converse is not necessarily valid because of the po[illegible] of cancellation). Hence by cofinality $y\times z$ does equal the [illegible] in the representation of yz, i.e. $y\times z = yz$. This [illegible]s the proof.

<u>Remark</u>. Note that in view of the above remark about the converse we do <u>not</u> have mutual cofinality. Fortunately, since the required inequality is trivially satisfied, we don't need it. At any rate, although very often it makes no difference, in general we must be careful as to which cofinality theorem is being used. For example, in the first part of the proof, we need mutual cofinality since otherwise we would require an inequality which is not at all obvious.

Theorems 5.7 and 5.8 give us a powerful tool for dealing with surreal numbers. In fact, for many purposes we can simply work with these generalized power series and ignore what surreal numbers are in the first place. This is an example of the whole spirit of abstraction in mathematics. However, there are limits to what can be accomplished by general power series methods since the surreal numbers are somewhat special. Note, for example, that the class of exponents is precisely the class of all surreal numbers, which in itself is unusual.

Let us see what the power series methods accomplish. First, we have an alternative way of dealing with inverses and square roots which is much easier than the direct method used in chapter three. Let us consider, for example, the inverse. The essential idea is as follows follows. Let $x = \omega^{a_0} r_0 \left(\sum_{i<\alpha} \omega^{b_i} s_i \right)$ where $b_i = a_i - a_0$ and $s_i = \frac{r_i}{r_0}$. Since the inverse of ω^{a_0} is ω^{-a_0} and since r_0, of course, has an inverse, it suffices to find the inverse of expressions of the form $\sum_{i<\alpha} \omega^{b_i} s_i$ where $b_0 = 0$ and $s_0 = 1$, i.e. of series which begin with 1. In fact, if $1 + \sum_{i<\alpha} \omega^{a_i} r_i$ is a series beginning with 1, we get the inverse by formally substituting $\sum_{i<\alpha} \omega^{a_i} r_i$ for x in $1-x+x^2-x^3 \cdots$. First, by lemma 5.1 this leads to a series which has meaning so that we obtain a surreal number. Then, theorems 5.7 and 5.8 guarantee that this is the inverse of $1+x$ since $(1-x+x^2-x^2\ldots)(1+x) = 1$ for ordinary formal series.

There is another method of using generalized power series to obtain existence results which does not depend on familiarity with identities for ordinary formal series. We shall apply this method to show that every positive surreal number has an nth root for any integer n. The same method can also be used to prove the existence of inverses. It is a generalization of the well-known procedure for ordinary formal power series $\sum_{i=0}^{\infty} a_i x^i$ where the coefficients of the various powers of x are obtained recursively. I like this method because of its elementary self-contained algebraic nature. We avoid any use of analysis and in particular the binomial theorem for fractional exponents.

Theorem 5.9. Every positive surreal number has an nth root for every positive integer n.

Proof. Let $\sum_{i<\alpha} \omega^{a_i} r_i$ be a surreal number. This can be expressed in the form $\omega^{a_0} r_0 [1+ \sum_{0<i<\alpha} \omega^{b_i} s_i]$. Now $\omega^{a_0} r_0$ has an nth root, namely $\omega^{a_0/n} \sqrt[n]{r_0}$, from theorem 5.4 and the fact that r_0 is positive. Hence it suffices to consider series which begin with 1.

Consider a series $1 + \sum_{i<\alpha} \omega^{a_i} r_i$. We shall express it in the form $(1+ \sum_{i<\beta} \omega^{x_i} y_i)^n$ by determining x_i and y_i inductively. (For ordinary power series $\alpha = \beta = \omega$ and the a_i's and x_i's are simply the integers. In our case the situation is slightly trickier. For example, β might be different from α.)

Suppose that $(1+ \sum_{i<\gamma} \omega^{x_i} y_i)^n$ agrees with $1 + \sum_{i<\alpha} \omega^{a_i} r_i$ for all terms ω^z where $z \geq x_i$ for some $i < \gamma$, but $(1+ \sum_{i<\gamma} \omega^{x_i} y_i)^n \neq 1 + \sum_{i<\alpha} \omega^{a_i} r_i$. (Recall that in our generalized power series the exponents are decreasing.) Then we claim that there exists x and y such that $x < x_i$ for all $i < \gamma$ and that $(1+ \sum_{i<\gamma} \omega^{x_i} y_i + \omega^x y)^n$ agrees with $1 + \sum_{i<\alpha} \omega^{a_i} r_i$ for all terms ω^x where $z \geq x$. Furthermore, if all x_i are finite linear combinations of the a_i with integral coefficients, then so is x. (The fact that x_i is not simply the same as a_i makes the process trickier than the one for ordinary power series.)

In fact, let x be the first exponent for which the coefficients of ω^x in $(1+ \sum_{i<\gamma} \omega^{x_i} y_i)^n$ and $1+ \sum_{i<\alpha} \omega^{a_i} r_i$ differ. Then $x < x_i$ for all i, and the respective coefficients, s and t, satisfy $s \neq t$. Note that s or t may be 0. Now consider an expression of the form $(1+ \sum_{i<\gamma} \omega^{x_i} y_i + \omega^x y)^n$. This agrees with $1+ \sum_{i<\gamma} \omega^{a_i} r_i$ for all terms ω^z where $z \geq x_i$ for some i. The earliest term for which there is possible disagreement is ω^x and, in fact, its coefficient is $s+ny$. Since $s \neq t$ there exists a non-zero y satisfying $s+ny = t$. (Uniqueness does not concern us.) With the above values for x and y the claim is clearly satisfied. Since x is either of the form a_i or a sum of x_i's, the second condition is clearly satisfied.

We now assume that $1 + \sum_{i<\alpha} \omega^{a_i} r_i$ does not have an nth root and obtain a contradiction. We use the claim above to define a sequence (x_i, y_i) inductively where i runs through the class of ordinals by letting $(x_\gamma y_\gamma)$ be the pair (x,y) obtained above. Since later terms have no effect on the coefficients of earlier exponents the induction works. However, since each x_i is a finite sum of a_i's, the collection of possible x_i's is a set so that eventually the sequence x_i must terminate. This contradiction proves the theorem.

Remark. It is interesting to compare this with a classical situation in which one is interested in power series which permit fractional exponents but only series of length ω. In that case one has the burden of showing that the sequence of exponents approaches ∞. Fortunately, we do not have this problem. For example, consider $1 + \omega^{-1} + \omega^{-\omega} + \omega^{-\omega-1} \ldots + \omega^{-\omega-n} \ldots$. This is a series of length ω. If we compute the square root using the proof of theorem 5.9 we begin with $1 + \frac{1}{2}\omega^{-1}$ and then obtain $1 + \frac{1}{2}\omega^{-1} - \frac{1}{8}\omega^{-2}$, etc. It is clear that we would need a sequence of length greater than ω. (For example, it it would take us "forever and a day" to reach the $\omega^{-\omega}$ term!) However, the proof shows that we must eventually terminate at some ordinal.

D APPLICATION TO REAL CLOSURE

We shall use the same technique as in the previous section to show that the class of surreal numbers forms a real closed field. Specifically, we adapt the classical Hensel's lemma argument to our transfinite series.

Lemma 5.7 (Variation on Hensel's lemma). Let $f(x) = x^n + \sum_{i=1}^{n} h_i x^{n-i}$ be a polynomial of degree n in the surreal numbers where h_i has the form $r_i + d_i$ with r_i real and d_i infinitesimal. (Thus all terms in the series expansion of h_i have non-positive exponents.) Suppose, furthermore, that $x^n + \sum_{i=1}^{n} r_i x^{n-i}$ factors into two relative prime polynomials P_0 and Q_0. Then $x^n + \sum_{i=1}^{n} h_i x^{n-i}$ factors into two polynomials, P and

Q, where P and Q have the same degrees as P_0 and Q_0 respectively and the first terms of the series expansions of the coefficients of P and Q are the same as the coefficients of P_0 and Q_0 respectively.

Proof. First, by regrouping we regard the polynomial f(x) as a series of the form $\sum_{i<\alpha} \omega^{a_i} s_i$ where s_i is a polynomial over the reals of degree at most n-1 for $i > 0$. Since a finite union of well-ordered sets is well-ordered, the a_i's are well ordered. By hypothesis $a_0 = 0$ and $s_0 = x^n + \sum_{i=1}^{n} r_i x^{n-i}$.

Let the degrees of P_0 and Q_0 be r and s respectively so $r+s = n$. We now extend P_0 and Q_0 in an inductive manner similar to that in our construction of nth roots. Suppose $(\sum_{i<\beta} \omega^{b_i} P_i)(\sum_{i<\beta} \omega^{b_i} Q_i)$ agrees with f(x) for all exponents y such that $y \geq b_i$ for some $i < \beta$ where the P_i's and Q_i's are polynomials of degrees at most r-1 and s-1 respectively for $i > 0$, but $(\sum_{i<\beta} \omega^{b_i} P_i)(\sum_{i<\beta} \omega^{b_i} Q_i) \neq f(x)$.

We shall find b_β such that $a_\beta < a_i$ for all $i < \beta$ and polynomials P_β and Q_β of degrees at most r-1 and s-1 respectively so that $(\sum_{i<\beta} \omega^{b_i} P_i + \omega^{b_\beta} P_\beta)(\sum_{i<\beta} \omega^{b_i} Q_i + \omega^{b_\beta} Q_\beta)$ agrees with f(x) for all exponents y such that $y \geq b_\beta$.

Let b_β be the first exponent x for which the coefficients of ω^x in $(\sum_{i<\beta} \omega^{b_i} P_i)(\sum_{i<\beta} \omega^{b_i} Q_i)$ and f(x) differ. Then $b_\beta < b_i$ for all i. Now consider the series $\sum_{i<\beta} \omega^{b_i} P_i + \omega^{b_\beta} G$ and $\sum_{i<\beta} \omega^{b_i} Q_i + \omega^{b_\beta} H$ where G and H are polynomials to be determined later. Then $(\sum_{i<\beta} \omega^{b} P_i + \omega^{b_\beta} G)(\sum_{i<\beta} \omega^{b_i} Q_i + \omega^{b_\beta} H)$ agrees with f(x) for all terms up to ω^{b_β}. The condition for agreement for the coefficients of ω^{b_β} is an equation of the form $HP_0 + GQ_0 = S$ for some polynomial S of degree at most n-1 because of the bounds on the degrees of P_i and Q_i. Since P_0 and Q_0 are relatively prime there exists G of degree at most r- and H of degree at most s-1 satisfying the above equation. Let

$P_\beta = G$ and $Q_\beta = H$. Also, as in the case of nth roots, if the b_i's are finite sums of the a_i then so is b_β. The rest of the argument is identical to the argument for n^{th} roots.

We regroup at the end so that we end up with monic polynomials of degree r and s over the surreals. (The justification for regrouping is the same as for ordinary polynomials in two variables. Of course the existence of bounds for the degrees of the polynomials is crucial for the regrouping to make sense.)

We are now ready for the main result of this section.

Theorem 5.10. Every polynomial equation of odd degree with surreal coefficients has a root. Furthermore the exponents which occur in the series expansion of the roots are rational linear combinations of the exponents which occur in the series expansions of the coefficients of the polynomial.

Proof. Let $P(x) = b_0x^n + b_1x^{n-1} + b_2x^{n-2} \ldots b_n$ be a polynomial of odd degree. We may assume that $b_0 = 1$ and $b_1 = 0$ by making the substitution $x = y - \frac{b_1}{n}$. The polynomial now has the form $x^n + \sum_{i=2}^{n} a_i x^{n-i}$.

Now suppose that the normal form of a_i begins with $\omega^{c_i} r_i$. Assume that the polynomial is not simply x^n. Let $c = \max_{i=2}^{n} \frac{c_i}{i}$. We now make the substitution $x = y\omega^c$. The equation becomes $(y\omega^c)^n + \sum_{i=2}^{n} a_i(y\omega^c)^{n-i} = 0$, which can be written in the form $y^n + \sum_{i=2}^{n} a_i\omega^{-ic}y^{n-i} = 0$. The coefficient of y^{n-i} begins with $(\omega^{c_i}r_i)\omega^{-ic}$. By choice of c we have $\frac{c_i}{i} \leq c$ with equality for at least one i, i.e. $c_i - ic \leq 0$. Thus all coefficients begin with terms with non-positive exponents and at least one term begins with exponent 0.

If an odd degree polynomial is factored into irreducible factors at least one of its factors must have odd degree. Hence to prove the theorem it is enough to show that an irreducible polynomial of odd degree must have degree one. If we apply the above construction to an irreducible polynomial the polynomial remains irreducible. Hence by the contrapositive of lemma 5.7 the real part of the polynomial does not have two relatively prime factors, i.e. has the form $(x-a)^n$ or $(x^2+bx+c)^n$. Since the degree is odd the latter possibility is ruled out. Hence the real part of the polynomial has the form $(x-a)^n$. Since the coefficient of x^{n-1} is 0, it follows that $a = 0$. Therefore the real part of the polynomial has the form x^n. This contradicts the fact that at least one term besides x^n begins with exponent 0.

Since the construction leads to a contradiction, the polynomial itself must be x^n. (We are not talking about the real part.) Since the polynomial is irreducible n must be 1.

The last part of the theorem follows from the same proof. For this purpose we restrict ourselves to surreal numbers whose exponents are of the form referred to in the statement of the theorem.

E SIGN SEQUENCE

Our aim in this section is to obtain a formula which expresses the sign sequence for $\sum_{i<\alpha} \omega^{a_i} r_i$ in terms of the sign sequences for a_i and r_i.

It is natural to look first for the sign sequence for ω^a. However, in order to carry through an induction we need to know the sign sequence for certain special finite sums along the way. Thus caution is required with the induction in order to avoid circular reasoning.

Specifically, we deal with finite sums of the form $\sum_{i<n} \omega^{a_i} r_i$, where a_{i+1} is an initial segment of a_i for all i and r_i is either an integer or a dyadic fraction with numerator 1. (It is understood that the a_i's are strictly decreasing, since we are working with normal forms.)

We first need some lemmas which are roughly variations on lemmas 5.2 and 5.3. In proving these lemmas we used the fact that $r = \{r-\varepsilon\}|\{r+\varepsilon\}$, which in the case where r is dyadic involves throwing out information. (This representation is cofinal but not mutually cofinal with the standard representation of r.) We now see what happens if we do not throw out information. In order to cut down on duplication later we shall deal with a general dyadic although for our immediate purpose we need only integers and dyadic fractions with numerator one.

Recall that by applying cofinality to the canonical representation of a dyadic fraction r we obtained r in the form $\{s\}|\{t\}$ if r is not an integer and in the form $\{r-1\}|\phi$ if r is a positive integer. For the special case where $r = \frac{1}{2^n}$ with $n > 0$ we have $\frac{1}{2^n} = \{0\}|\{\frac{1}{2^{n-1}}\}$.

Lemma 5.8. (a) If $r = \{r'\}|\{r''\}$ and the set of lower elements of a is non-empty, then $\omega^a r = \{\omega^a r' + \omega^{a'} n\}|\{\omega^a r'' - \omega^{a'} n\}$ where a' as usual is a typical lower element in the canonical representation of a and n is an arbitrary positive integer. If the set is empty then $\omega^a r = \{\omega^a r'\}|\{\omega^a r''\}$.

b) If n is a positive integer greater than one then $\omega^a n = \{\omega^a(n-1) + \omega^{a'} m\}|\{\omega^{a''} \varepsilon\}$ if the set of lower elements of a is non-empty, where m is an arbitrary positive integer, a' is as before, a'' is as usual a typical upper element in the canonical representation of a, and ε is an arbitrary positive dyadic fraction with numerator 1. If the set is empty then $\omega^a n$ is $\{\omega^a(n-1)\}|\{\omega^{a''} \varepsilon\}$.

Proof. (a) We compute $\omega^a r$ as in the proof of lemma 5.2. Again $\omega^a = \{0, s\omega^{a'}\}|\{t\omega^{a''}\}$. Hence $\omega^a r$ is $\{\omega^a r', \omega^a r' + (s\omega^{a'})(r-r'), \omega^a r'' - (t\omega^{a''})(r''-r)\}|\{\omega^a r'', \omega^a r'' - (s\omega^{a'})(r''-r), \omega^a r' + (t\omega^{a''})(r-r')\}$. By cofinality we can eliminate the third terms among the upper and lower elements and replace terms such as $\omega^a r' + (s\omega^{a'})(r-r')$ by $\omega^a r' + \omega^{a'} m$. Also if the set of a' is non-empty we can eliminate the first terms by cofinality. Thus we get the desired form.

(b) In this case there is no r''. The computation is the same except that now we need the third term of the upper elements since the other terms are not present. Since $\omega^{a''} \gg \omega^a$, by cofinality this term may be replaced by $\omega^{a''} \varepsilon$.

Note that negative integers can be handled by sign reversal. Also, $r' = 0$ for dyadic fractions with numerator one so that the formulas simplify. If the set of a's is non-empty then a typical lower element is $\omega^{a'} n$. Otherwise 0 is the only lower element.

We are now ready to consider finite sums. A convenient representation for n-fold sums has already been mentioned in Chapter 3.

In fact, $\sum_{i=1}^{n} a_i$ can be expressed as

$\{a_1+a_2...+a_j{}'...+a_n\}|\{a_1+a_2...+a_k{}''...a_n\}$ where $1 \leq j \leq n$ and $1 \leq k \leq n$. We now apply this to sums of the form $\sum_{i<n} \omega^{a_i} r_i$ referred to earlier and use cofinality to simplify. It is understood that $r_i \neq 0$ for all i.

<u>Lemma 5.9</u>. Let $\sum_{i<n} \omega^{a_i} r_i$ be a surreal number where for all i, a_{i+1} is an initial segment of a_i less than a_i and r_i is a dyadic fraction. Then $\sum_{i<n} \omega^{a_i} r_i$ can be expressed in the form $F|G$ where a typical element x in G is obtained as follows:

(a) If r_{n-1} is not a positive integer let $r_{n-1}{}''$ be the minimum upper element in the canonical representation of r_{n-1}. Then $x = \sum_{i<n-1} \omega^{a_i} r_i + \omega^{a_{n-1}} r_{n-1}{}'' - \omega^{a_{n-1}{}'} m$ where $a_{n-1}{}'$ is a typical lower element in the canonical representation of a_{n-1} and m is an arbitrary positive integer. (If there is no $a_{n-1}{}'$ the last term is omitted.)

(b) If r_{n-1} is a positive integer but r_i is not a positive integer for at least one i, then let j be the largest index for which r_j is not a positive integer. Then $x = \sum_{i<j} \omega^{a_i} r_i + \omega^{a_j} r_j{}'' - \omega^{a_j{}'} m$ where $r_j{}''$ is the minimum upper element in the canonical representation of r_j, $a_j{}'$ is a typical lower element in the canonical representation of a_j and m is an arbitrary positive integer.

(c) If r_i is a positive integer for all i and a_0 is not an ordinal then $x = \omega^{a_0{}''} \delta$ where $a_0{}''$ is a typical upper element in the canonical representation of a_0 and δ is a positive dyadic fraction with numerator one.

(d) If r_i is a positive integer for all i and a_0 is an ordinal then G is empty. (In fact, such an x is clearly an ordinal.)

A typical element of F is obtained similarly.

<u>Proof</u>. This follows easily from lemma 5.8 and the representation of n-fold sums discussed earlier using cofinality. Specifically, in the

expression $\sum_{i<n} \omega^{a_i}r_i$ we replace $\omega^{a_k}r$ by x for some k where x is an element in the representation of $\omega^{a_k}r_k$ given by lemma 5.8. We consider G. F can be handled similarly.

In part (a) the situation is analogous to that of lemma 5.3, i.e. by the lexicographical order, the terms obtained by replacing $\omega^{a_{n-1}}r_{n-1}$ by x are cofinal. This gives us the result. For part (b) let us analyze more explicitly what happens when $\omega^{a_k}r_k$ is replaced by x. If r_k is not a positive integer $\sum_{i<n} \omega^{a_i}r_i$ is replaced by $\sum_{i<k} \omega^{a_i}r_i + \omega^{a_k}r_k'' - \omega^{a_k'}m + \sum_{k<i<m} \omega^{a_i}r_i$. In particular, the first term altered is the ω^{a_k} term. If r_k is a positive integer the sum is replaced by $\sum_{i<k} \omega^{a_i}r_i + \omega^{a_k''}\delta + \sum_{k<i<n} \omega^{a_i}r_i$. The higher term $\omega^{a_k''}$ is introduced. Thus in either case when $\omega^{a_k}r_k$ is replaced by x the term that is altered is at least the ω^{a_k} term.

Now if $k = j$, it follows that the ω^{a_j} term is the first one that is altered. For $k < j$ the term that is altered is necessarily higher (either ω^{a_k} or still higher). Thus the terms obtained by altering $\omega^{a_j}r_j$ are cofinal with respect to the terms obtained by altering $\omega^{a_k}r_k$ for $k < j$. (So far this may be regarded as a more detailed proof of part (a) if j is replaced by n.) Now suppose $k > j$. Recall that a_k'' is an initial segment of a_k and $a_k'' > a_k$. Since a_k is an initial segment of a_j so is a_k''. Furthermore $a_k'' > a_j$. Since $a_k < a_j$ we appear to have an inequality going in the wrong direction to apply transitivity. Nevertheless the inequality follows from the nature of initial segments. Suppose a_k'' has length α. Since $a_k'' > a_k$, we have of $a_k(\alpha) = -$. Since a_k is an initial segment of a_j, then $a_j(\alpha) = -$. Hence $a_k'' > a_j$. Thus the terms obtained by altering $\omega^{a_j}r_j$ are cofinal with respect to the terms obtained by altering $\omega^{a_k}r_k$ for $k > j$.

We have shown that the upper terms may be simplified to $\sum_{i<j} \omega^{a_i}r_i + \omega^{a_j}r_j'' - \omega^{a_j'}m + \sum_{j<i<n} \omega^{a_i}r_i$, i.e. by cofinality the only term we need alter is the jth term.

If $b = \max(a_j', a_{j+})$ then for m' sufficiently high $\omega^b m' - \omega^{a_j'} m > |\sum_{j<i<n} \omega^{a_i} r_i|$. Hence it is clear by mutual cofinality that the above expression can be simplified to

$\sum_{i<k} \omega^{a_i} r_i + \omega^{a_j''} r_j'' - \omega^{a_j'} m$. This completes the proof of part (b).

In part (c) all sums obtained have the form

$\sum_{i<k} \omega^{a_i} r_i + \omega^{a_k''} \delta + \sum_{k<i<n} \omega^{a_i} r_i$. It is standard by now that this is mutually cofinal with $\sum_{i<k} \omega^{a_i} r_i + \omega^{a_k''} \delta$. During our proof of part (b) we already saw that a_k'' is an initial segment of a_0 and that $a_k'' > a_0$. Hence the above expression is mutually cofinal with $\omega^{a_k''} \delta$. When $k = 0$ this is simply $\omega^{a_0''} \delta$. Finally, since every element of the form a_k'' is also of the form a_0'', we need only terms for which $k = 0$, thus completing the proof of part (c).

Part (d) follows immediately since there are no a_0''. (Recall that a_0 being an ordinal is equivalent to a_0 consisting only of pluses which is in turn equivalent to the non-existence of any a_0''.)

It is interesting to contrast the situations in parts (a) and (c) with regard to cofinality. In part (a) the last term contributes the cofinal part whereas the reverse is true in part (c).

We are now ready to determine the formula for the sign sequence for ω^a. Let a_α be the number of pluses in the initial segment of a of length α, and let a^+ be the total number of pluses in a.

<u>Theorem 5.11.</u> (a) The sign sequence of ω^a is as follows. We begin with a plus. Then for each α we have a string of $\omega^{a_\alpha+1}$ pluses if $a(\alpha) = +$ and $\omega^{a_\alpha+1}$ minuses if $a(\alpha) = -$.

(b) The sign sequence of $\omega^a n$, where n is a positive integer greater than 1 is obtained by beginning with the sequence for ω^a and following it by $\omega^{a^+}(n-1)$ pluses. The sign sequence of $\omega^a \frac{1}{2^n}$ where n is a positive integer is obtained by beginning with the sequence for ω^a and following it by by $\omega^{a^+} n$ minuses. For negative coefficients we use sign reversal. (Note that we still count the pluses in a since a is unaltered.)

(c) The sign sequence of $\sum_{i<n} \omega^{a_i} r_i$, where r_i is either an integer or a dyadic fraction with numerator one, and, for all i, a_{i+1} is an initial segment of a_i, is obtained by juxtaposing all the modified sign sequences for each i, where the modified sign sequence of $\omega^{a_i} r_i$ is obtained as follows: For $i = 0$ we use the rule in part (b). For $i > 0$ we apply rule (b) to the element $\omega^{b_i} r_i$, where b_i is obtained from a_i by ignoring all minuses.

Before proving the theorem we illustrate with several examples. First let $a = (+-)$. Then ω^a begins with a plus. Then the first term in a, which is $+$, gives rise to $\omega^{0+1} = \omega$ pluses and the second term, which is $-$, gives rise to $\omega^{1+1} = \omega^2$ minuses. So altogether we have $1+\omega = \omega$ pluses followed by ω^2 minuses. (In juxtaposing sequences ordinal addition is what is relevant.) Incidentally, since $a = \frac{1}{2}$ this is $\sqrt{\omega}$ (by the law of exponents), so that this is consistent with an example which was done in Chapter 4.

Now let $a_0 = (-+-++)$, $a_1 = (-+-+)$ and $a_2 = (-+-)$. We compute $\omega^{a_0}5 + \omega^{a_1}\frac{1}{4} - \omega^{a_2}3$. By rule (a) we begin with a plus, ω minuses, ω pluses, ω^2 minuses, ω^2 pluses, and finally ω^3 pluses. (The group of ω^2 pluses gets absorbed by the ω^3 pluses.) By rule (b) this is followed by $\omega^3 4$ pluses contributed by the "5". The contribution from $\omega^{a_1}\frac{1}{4}$ follows. By rule (c) this is the sequence obtained from $\omega^{b_1}\frac{1}{4}$ where $b_1 = (++)$. This consists of ω^2 pluses followed by $\omega^2 2$ minuses. (Note that the contribution of the "$\frac{1}{4}$" is the same for $\omega^{a_1}\frac{1}{4}$ and $\omega^{b_1}\frac{1}{4}$.) Finally we have $\omega \cdot 3$ minuses because of sign reversal since we would have had $\omega \cdot 3$ pluses if the term was $+\omega^{a_2}3$.

The example suggests that the formula can be simplified if we consider blocks of pluses and minuses in a surreal number a rather than individual signs. In fact, this can and will be done later. However, the present form is appropriate for the inductive proof.

Proof (of Theorem 5.11). We do this by induction on the length of the sign sequence $g(x)$ obtained from $x = \sum_{i<n} \omega^{a_i} r_i$ by the formula in the statement of the theorem. We want to prove that $g(x) = x$.

First we show that g is one-one. Suppose $x = \sum_{i<m} \omega^{a_i} r_i$, $y = \sum_{i<n} \omega^{b_i} s_i$, and $g(x) = g(y)$. Assume first that $a_0 = b_0$. Then $g(x)$ and $g(y)$ have the same tail after discarding the initial segments corresponding to $\omega^{a_0} = \omega^{b_0}$. Now $a_0^+, a_1^+, \ldots, a_{m-1}^+$ is a decreasing sequence of ordinals since $a_{i+1} < a_i$ and a_{i+1} is an initial segment of a_i. Hence the length of the tail has the form $\sum_{i<m} \omega^{a_i^+} n_i$.

Thus the length of the tail determines a_i^+ and n_i uniquely. Furthermore, a_i^+ determines a_i uniquely. This is because a_i is obtained from a_0 by stopping at a plus. Finally, the signs of the various strings determine whether r_i is an integer or a dyadic with numerator 1 and the value of n_i determines r_i. Thus $x = y$.

Now we rule out the possibility that $a_0 \neq b_0$. If neither a_0 nor b_0 is an initial segment of the other, then clearly a discrepancy between $g(x)$ and $g(y)$ arises at the point where a_0 and b_0 differ. Suppose without loss of generality that a_0 is an initial segment of b_0, and consider the tail following the sequence for ω^{a_0}. The length of the tail of $g(x)$ has the form $\sum_{i<m} \omega^{a_i^+} n_i$ which is less than $\omega^{a_0^+ + 1}$. The tail of $g(x)$ begins with a string of $\omega^{a_0^+ + 1}$ identical signs. So certainly $g(x) \neq g(y)$.

It is clear that $g(x)$ has finite length only if x has the form $\omega^0 r = r$, in which case the formula is consistent with what we already know about dyadic fractions.

Now let $x = \omega^a$. We look at the canonical representation $g(x) = F|G$. An element of F is obtained by stopping just before a plus in the sequence for $g(x)$. If a consists only of minuses then the only plus in x is at the beginning so $F = \{0\}$. Otherwise a plus in x rises from a plus in a at some place α where $a(\alpha) = +$. Let b be the initial segment of a of length α. There are $\omega^{b^+ + 1}$ pluses in x arising from that plus in a. Then the typical element of F is obtained by juxtaposition of the sequence arising from b with c pluses where $c < \omega^{b^+ + 1}$. By cofinality we may just as well limit ourselves to values of c of the form $\omega^{b^+} n$ for positive integers n. But by case (b) and the inductive hypothesis, such an element is $\omega^b(n+1)$.

Similarly an element of G is obtained by stopping just before a minus in the sequence for $g(x)$. Let b be an initial segment of a of length α where $a(\alpha) = -$. Then (again using cofinality) we obtain the typical element of G by juxtaposition of the sequence arising from b $\omega^{b^+}n$ minuses. By the inductive hypothesis this is $\omega^b(\frac{1}{2^n})$. Therefore $g(x) = \{0,\omega^{a'}n\}|\{\omega^{a''}(\frac{1}{2^n})\} = \omega^a = x$.

Next let $x = \omega^a n$, where n is a positive integer larger than one. Since $g(x)$ is obtained from $g(\omega^a)$ by adding pluses only, both have the same upper elements. For the lower elements we may limit ourselves to the contribution of the term "n" by cofinality. Thus we add on c pluses to the sequence for ω^a, where $c < \omega^{a^+}(n-1)$. Again by cofinality we assume that c has the form $\omega^{a^+}(n-2) + \omega^b m$ where b is an ordinal less than a^+ and m is a positive integer. Now as a' runs through all the initial segments of a less than a, a'^+ runs through all ordinals less than a^+, so c has the form $\omega^{a^+}(n-2) + \omega^{a'^+}m$. But this is exactly $g(\omega^a(n-1) + \omega^{a'}m) = \omega^a(n-1) + \omega^{a'}m$ by the inductive hypothesis. (If there are no terms a' this reduces to $\omega^a(n-1)$.) In any case the upper and lower elements for $g(x)$ are just what we need by lemma 5.8(b) to deduce that $g(x) = \omega^a n = x$.

Now let $x = \omega^a(\frac{1}{2^n})$. Since this is similar to the previous case it suffices to outline the argument. $g(x)$ has the same lower elements as $g(\omega^a) = \omega^a$. The upper elements are obtained by adding on $\omega^{a^+}(n-1) + \omega^{a'^+}m$ minuses. By the inductive hypothesis this is $\omega^a(\frac{1}{2^{n-1}}) - \omega^{a'}m$. Since $\frac{1}{2^n} = \{0\}|(\frac{1}{2^{n-1}})$ we have just what we need by lemma 5.8(a) to deduce that $g(x) = \omega^a(\frac{1}{2^n}) = x$.

We now let $x = \sum_{i<n} \omega^{a_i}r_i$. The argument is a straightforward application of lemma 5.9. Let $g(x) = F|G$ be the canonical representation of $g(x)$. Suppose first that r is not a positive integer but r_i is a positive integer for $i > j$. In order to get a set G' cofinal in G, it suffices to consider a set of minuses in $g(x)$ which is arbitrarily far out. We obtain this from the contribution of the term $\omega^{a_j}r_j$. Thus a typical element of G' is obtained by juxtaposing the sequence obtained from the truncated sum up to the jth term with a typical upper element in the canonical representation of $g(\omega^{b_j}r_j)$, where

b_j is obtained from a_j as in the statement of part (c) of the theorem. In other words, the typical element of G' has the form $g(\sum_{i<j} \omega^{a_i} r_i + \omega^{a_j} r_j'' - \omega^{a_j'} m)$ where r_j'', a_j', and m are as in the statement of lemma 5.9(b). This follows by the earlier part of the proof dealing with monomials. By the inductive hypothesis this is $\sum_{i<j} \omega^{a_i} r_i + \omega^{a_j} r_j'' - \omega^{a_j'} m$.

If r_i is a positive integer for all i, then g(x) is obtained from $g(\omega^{a_0})$ by adding on pluses only. Hence g(x) and $g(\omega^{a_0})$ have the same upper elements, namely $\omega^{a_0''} \varepsilon$.

A similar argument applies to the lower elements. In all cases the upper and lower elements obtained for g(x) are just what we need by lemma 5.9 to deduce that g(x) = x. (For convenience we unified various cases. For example, in lemma 5.9 part (a) may be regarded as a special case of part (b) and part (d) of part (c). There is a pedagogical advantage in separating cases at the beginning for the sake of concreteness, but at a later stage it is repetitious and tedious.)

We are now ready to determine the sign sequence of a general sum $\sum_{i<\alpha} \omega^{a_i} r_i$. First, we define what we mean by a reduced sequence a_β' of a_β, where $\beta < \alpha$.

The reduced sequence a_β^{o} of a_β is obtained from a_β by discarding the following minuses occurring in a_β:

I if $a_\beta(\delta) = -$ and there exists $\gamma < \beta$ such that $(\forall x \leq \delta)[a_\gamma(x) = a_\beta(x)]$, then the δth minus is discarded

II if β is a non-limit ordinal, a_β has $a_{\beta-1}$ followed by a minus as an initial segment, and $r_{\beta-1}$ is not dyadic, then the last minus is discarded.

For example, if $a_0 = (+-++)$ and $a_1 = (+-+-)$ then $a_1^{0} = (++-)$, i.e. the first minus is discarded but not the second one.

If $a_0 = (+++)$ and $a_1 = (+++-)$ then $a_1^{0} = (+++)$ if r_0 is not dyadic but $a_1' = a_1$ if r_0 is a dyadic fraction.

Roughly speaking, we ignore minuses which occur earlier; however, the second part gives a special situation where even a new minus is ignored.

Theorem 5.12. (a) The sign sequence for $\omega^a r$ for positive real r is obtained by juxtaposing the sequence for ω^a with the sequence obtained from r by omitting the first plus and repeating each sign in r, ω^{a^+} times.

(b) If r is negative the sign sequence in (a) is reversed.

(c) The sign sequence for $\sum_{i<\alpha} \omega^{a_i} r_i$ is obtained by juxtaposition of the sign sequence for the successive $\omega^{a_i^o} r_i$ where a_i^o is the reduced sequence of a_i.

Remarks. Theorem 5.11 is, of course, a special case of theorem 5.12. In fact, recall that in theorem 5.11 we were interested primarily in the sign sequence for ω^a and therefore used only those sums which were needed for the induction.

We illustrate theorem 5.12(a) with a simple example. Consider $\omega^{\frac{1}{2}}(\frac{3}{4})$. Here $a = (+-)$ and $r = (+-+)$. We already saw earlier that $\omega^{\frac{1}{2}}$ gives rise to ω pluses followed by ω^2 minuses. Since $a^+ = 1$ the contribution from r is ω minuses followed by ω pluses.

As a simple example of theorem 5.12(c) consider $\omega^{\frac{1}{2}} + \omega^{\frac{1}{8}}$. $\frac{1}{8} = (+---)$. Since $\frac{1}{2} = (+-)$ we ignore the first minus in determining the contribution of the term $\omega^{\frac{1}{8}}$. This therefore becomes ω pluses followed by $\omega^2 2$ minuses.

Proof. We first consider the case where the surreal number has the form $\sum_{i<n} \omega^{a_i} r_i$ with r_i arbitrary dyadic. This is a slight generalization of theorem 5.11. The proof that g is one-one extends immediately to this case. For example, the signs of the strings still determine r_i uniquely. Recall also that lemmas 5.8 and 5.9 deal with general dyadic coefficients. This gives us a head start in imitating the proof of theorem 5.11.

The subcase where $x = \omega^a r$ with r a positive dyadic fraction but neither an integer nor a dyadic fraction with numerator one is similar to the cases $x = \omega^a n$ and $x = \omega^a(\frac{1}{2^n})$ dealt with in the proof of theorem 5.11. In fact, let $r = \{r'\}|\{r''\}$. Note that r' is the initial segment of r obtained by stopping just before the last plus and r'' by stopping just before the last minus. By hypothesis r begins with a plus following which there is at least one plus and one minus. Hence in the canonical representation of $g(x)$ we obtain cofinality on both sides by limiting ourselves to the contribution of the term "r". By further use of cofinality a typical upper element is obtained by adding on $\omega^{a'+}m$ minuses to $g(\omega^a r'')$ and similarly for typical lower elements. By the inductive hypothesis such a typical upper element is $\omega^a r' - \omega^{a'} m$. This and a similar result for lower elements gives us what we need by lemma 5.8(a) to deduce that $g(x) = \omega^a r = x$.

For finite sums the proof is identical to that of theorem 5.11(c) since no use is made there of the assumption that the dyadics r_i have numerator one.

Next, we consider the case $x = \omega^a r$ where r is not dyadic. The last part of the proof that g is one-one, which depends only on length, no longer works. For example, $g(\omega^2)$ and $g(\omega\sqrt{2})$ have the same length ω^2 by the formula. (Of course, if one is interested, the proof can be extended to the present case by noting that even if the length of the tail of $g(x)$ is $\omega^{a_0^+ +1}$ the signs are necessarily not all alike so $g(x) \neq g(y)$. On the other hand, this is not too important now since it is no longer urgent to know in advance that g is one-one.)

Let $r = R'|R''$ be the canonical representation. By lemma 5.2 $\omega^a r$ may be expressed as $\omega^a R' | \omega^a R''$ since r is not dyadic. (Note the crucial simplification for non-dyadic coefficients where we get by with lemma 5.2 rather than lemma 5.8(a)). Consider the lower elements in the canonical representation of $g(x)$. Since r does not have a last plus we obtain a cofinal subset by taking only those initial segments which stop just before a string of pluses which correspond to a plus in r. But this is precisely of the form $g(\omega^a r') = \omega^a r'$ since we have the result for dyadic coefficients. Similarly a typical upper element has the form $(\omega^a r'')$. Hence $g(x) = \omega^a r' | \omega^a r'' = \omega^a r = x$.

The fact that it is possible for x to be an initial segment of y and still have $n\ell(x) > n\ell(y)$ has helped to complicate the proof so far. We were unable to use induction on $n\ell(x)$ which a priori seems like the natural type of induction to use. Fortunately, for the rest of the proof we can use induction on a quantity closely related to $n\ell(x)$

instead of the length of $g(x)$.

We now define $rn\ell(x)$ (the reduced normal length of $g(x)$.

Let $\sum_{i<\alpha} \omega^{a_i} r_i$ be the normal form of x. First we define $h(i)$ for $i < \alpha$.

(a) If r_i is not dyadic then $h(i) = i+1$ providing $i+1 < \alpha$.

(b) If r_i is dyadic then $h(i)$ is the least j exceeding i such that either r_j is not dyadic or $\exists k(i \leq k < j)$ and a_j is not an initial segment of a_k) providing $j < \alpha$]. $h(i)$ may be undefined for some i.

We now obtain a subsequence d_i of α as follows:

(a) $d_0 = 0$.

(b) $d_{i+1} = h(d_i)$.

(c) If β is a limit ordinal then $d_\beta = \lim_{\gamma<\beta} d_\gamma$ providing $\lim_{\gamma<\beta} d_\gamma < \alpha$.

Finally, $rn\ell(x)$ is the ordinal type of the sequence $\{d_i\}$ (i.e the least i:d_i is undefined).

Note that the definition says that if $h(i) = j$ then for $i \leq k < k+1 < j$, a_{k+1} is an initial segment of a_k. Hence necessarily $j = i+n$ for a <u>finite</u> n. Furthermore $\sum_{i\leqslant k<j} \omega^{a_k} r_k$ is a sum of the kind we considered earlier since again by the definition all r_k are necessarily dyadic if the sum does not reduce to a monomial. Thus, informally, we obtain the reduced normal length of x by regarding all such finite blocks in the expansion of x as a single term.

We now need a lemma which bears the same relation to lemma 5.9 that lemma 5.3 bears to lemma 5.2. Let x have the form $\sum_{i<\alpha+n} \omega^{a_i} r_i$ where for all $i \geq \alpha$, a_{i+1} is an initial segment of a_i and r_i is dyadic. I.e. x has the form $\sum_{i<\alpha} \omega^{a_i} r_i + y$ where y is a finite sum of the kind considered earlier. Note that $rn\ell(x) \leq rn\ell(\sum_{i<\alpha} \omega^{a_i} r_i) + 1$.

<u>Lemma 5.10</u>. Let x have the normal form $\sum_{i<\alpha+n} \omega^{a_i} r_i$ and express x as $\sum_{i<\alpha} \omega^{a_i} r_i + y$. Suppose that y is a surreal number which satisfies the hypothesis of lemma 5.9. Then x can be expressed in the form $F|G$

where a typical element x'' of G is obtained as follows:

(1) If y satisfies case (a) or (b) of lemma 5.9 then

$$x'' = \sum_{i<\alpha} w^{a_i} r_i + y'' \quad \text{where } y'' \text{ is as in the lemma.}$$

(2) If y satisfies case (c) and $(\exists a_\alpha'')(\forall_i(i<\alpha \rightarrow a_\alpha''<a_i)$ then

$$x'' = \sum_{i<\alpha} \omega^{a_i} r_i + \omega^{a_\alpha''} \delta.$$

(3) If y satisfies case (c) and no such a_α'' exists or if y satisfies case (d) (in which case there certainly does not exist an a_α'') then x'' is a typical upper element in the representation of $\sum_{i<\alpha} \omega^{a_i} r_i$ as given by the definition if α is a limit ordinal and by lemma 5.3 if α is a non-limit ordinal.

A typical element of F is obtained similarly.

Proof of lemma. Typical elements x'' are $\sum_{i<\alpha} \omega^{a_i} r_i + y''$ and $z''+y$ where z'' is a typical upper element of $\sum_{i<\alpha} \omega^{a_i} r_i$. In cases (1) and (2) terms of the former type are clearly cofinal by the lexicographical order thus proving the lemma in these cases. In case (3) consider a typical element of the former type. This has the form $\sum_{i<\alpha} \omega^{a_i} r_i + \omega^{a_\alpha''} \varepsilon$.

Then by definition of case (3) $(\exists j)(j < \alpha \wedge a_j \leq a_\alpha'')$. Now $\sum_{i<j} \omega^{a_i} r_i + \omega^{a_j} \varepsilon'$ is clearly less than $\sum_{i<\alpha} \omega^{a_i} r_i + \omega^{a_\alpha''} \varepsilon$ for $\varepsilon' < \varepsilon$ by the lexicographical order. This guarantees that terms of the type $z'' + y$ are cofinal in this case. As usual, by cofinality such terms may be replaced by z''. This completes the proof of the lemma.

The distinction between case (2) and case (3) can be expressed in terms of the sign sequences of the a_i's. Recall that case (3) is characterized by the condition $(\forall a_\alpha'')(\exists a_j)(j<\alpha \quad a_j \leq a_\alpha'')$. Since the a_i's are decreasing and since the a_α'' are initial segments corresponding to minuses in a_α, an a_j corresponds to a_α'' by the above condition precisely if it either equals a_α'' or has a_α'' followed by a minus as an initial segment. Furthermore, the existence of an a_j satisfying $a_j < a_\alpha''$ for a given a_α'' is precisely condition I in the definition of reduced sequence for discarding minuses for the minus a_α corresponding to this a_α''.

Let us maintain the condition for case (3) but assume that for some a_α'' there is no a_j satisfying $a_j < a_\alpha''$. Then the corresponding a_j satisfies $a_j = a_\alpha''$. Since there is no a_j satisfying $a_j < a_\alpha''$, j must be an immediate predecessor of α, i.e $\alpha = j+1$. This is condition (II) in the definition of reduced sequence except for the lack of reference to the nature of $r_{\alpha-1}$. Furthermore the minus corresponding to a_α'' must be the last minus in a_α.

We now have what we need for the main induction on $rn\ell(x)$. First suppose that $rn\ell(x)$ is a non-limit ordinal. Let $x = \sum_{i<\alpha+n} \omega^{a_i} r_i$ where $[(\forall i \geq \alpha)(a_{i+1}$ is an initial segment of a_i and r_i is dyadic) or ($n=1$ and r_i is non-dyadic)$]$. Then $x = \sum_{i<\alpha} \omega^{a_i} r_i + y$ where y is a finite sum of the kind considered earlier. By the inductive hypothesis we may assume the result for $\sum_{i<\alpha} \omega^{a_i} r_i$. We now use induction on $\ell g(y^o)$ where $y^o = \sum_{\alpha \leq i < \alpha+n} \omega^{a_i^o} r_i$. The argument is similar to the one used in the proof of theorem 5.11; however, there is a complication because of the need to consider reduced sequences. In this connection note the obvious fact concerning juxtaposition of sequences that if $A = F|G$ where F and G are both non-empty then $SA = SF|SG$.

As before, we desire to show that g is one-one, but now we regard g as a function of y using the reduced sequence for fixed $\sum_{i<\alpha} \omega^{a_i} r_i$. This can differ from the earlier case only by the contribution of a_α to the sign sequence since all minuses occuring in a_i for $i > \alpha$ are automatically ignored. Thus it suffices to show that $a_\alpha \to a_\alpha^o$ is one-one. The immediate reaction may be that it is unreasonable to expect this but recall that $(\forall i<\alpha)(a_i > a_\alpha)$ so that the apparently obvious way of getting counter-examples fails.

Specifically, suppose $a_\alpha \neq b_\alpha$ but $a_\alpha^o = b_\alpha^o$. Assume $a_\alpha(j) \neq b_\alpha(j)$ but $a_\alpha(i) = b_\alpha(i)$ for $i < j$. Without loss of generality the "dangerous" case occurs if $a_\alpha(j) =$ minus and the minus is ignored in a_α^o. Hence $(\exists \beta<\alpha)(\forall k<j)[a_\beta(k) = a_\alpha(k)]$. By the lexicographical order, since $b_\alpha < b_\beta = a_\beta$, $b_\alpha(j)$ is necessarily minus regardless of whether condition I or II holds in the definition of reduced sequence, thus leading to a contradiction.

Now the sign sequence for $g(\sum_{i<\alpha+n} \omega^{a_i} r_i)$ is the juxtaposition $[g(\sum_{i<\alpha} \omega^{a_i} r_i)][g(y^o)]$. (Note that we already know by theorem 5.11 that $g(y) = y$ and similarly for y', y^o, etc. but it is convenient to maintain the notation $g(y)$ for consistency of notation in dealing with juxtaposition.

Suppose y is positive. A similar argument will apply if y is negative. Suppose first that $g(y^o)$ contains a minus not contributed by a_α. This corresponds to y as in lemma 5.10(1). We can obtain subsets cofinal in the canonical representation of $g(x)$ by considering pluses and minuses in the segment within $g(y^o)$. Now if $y = \{y'\}|\{y''\}$ then $y^o = \{y'^o\}|\{y''^o\}$ and furthermore

$$[g(\sum_{i<\alpha} \omega^{a_i} r_i)][g(y^o)] = \{[g(\sum_{i<\alpha} \omega^{a_i} r_i)][g(y'^o)]\}|\{[g(\sum_{i<\alpha} \omega^{a_i} r_i)][g(y''^o)]\}$$

by trivial reasoning with juxtaposition.

By the inductive hypothesis this is

$\{\sum_{i<\alpha} \omega^{a_i} r_i + y'\}|\{\sum_{i<\alpha} \omega^{a_i} r_i + y''\}$ which is $\sum_{i<\alpha} \omega^{a_i} r_i + y$ as desired by lemma 5.10(1). It is worth remarking that juxtaposition works trivially in the argument here since the minuses ignored in y depend only on the nature of a_i for $i < \alpha$ so they are the same for all y' and y''.

Now suppose all minuses in $g(y^o)$ are contributed by a_α, i.e. by minuses in a_α which are not ignored. Such a minus corresponds to an a_α''. Assume first that case 2 of lemma 5.10 holds. Then by cofinality we may limit ourselves to such a_α'' and the juxtaposition argument is identical to that of the earlier case.

The most subtle case occurs when case 2 is not satisfied. By the remarks following the proof of lemma 5.10 this can happen only when α has the form $j+1$ and the minus corresponding to $a_\alpha'' = a_j$ is the last minus in a_α. Since we are dealing with a minus in α which is not ignored, $r_{\alpha-1}$ is dyadic. (This case is the most delicate with regard to the issue of ignoring of minuses.) We obtain a subset cofinal in the upper elements of the canonical representation of $g(x)$ by taking sequences of the form $g(\sum_{i<\alpha} \omega^{a_i} r_i)$ followed by $\omega^{a_\alpha^+}$ pluses and $\omega^{a_\alpha^+} n$ minuses for some integer n. This follows because first all minuses preceding the one corresponding to a_α'' are ignored in y^o.

Next it follows from the identity $\sum_{i<\alpha} \omega^{i+1} = \omega^{\alpha}$ that $g(y^{o})$ begins with $\omega^{a^{+}}$ pluses.

Finally, there are no minuses following the succeeding $\omega^{a_{\alpha}^{+}+1}$ minuses contributed by the minus corresponding to $a_{\alpha}"$. Now $\sum_{i<\alpha} \omega^{a_i} r_i = \sum_{i<j} \omega^{a_i} r_i + \omega^{a_j} r_j$. Also the tail of $g(\sum_{i<\alpha} \omega^{a_i} r_i)$ contributed by $\omega^{a_j} r_j$ followed by the $\omega^{a_j^{+}}$ pluses and $\omega^{a_j^{+}} n$ minuses is exactly the contribution of a term of the form $\omega^{a_j}(r_j+\varepsilon)$ for some positive dyadic ε to $g[\sum_{i<j} \omega^{a_i} r_i + \omega^{a_j}(r_j+\varepsilon)]$, which is $\sum_{i<\alpha} \omega^{a_i} r_i + \omega^{a_j}\varepsilon$ by the main inductive hypothesis (the one on $rn\ell(x)$). Hence $g(x)$ has the form $\{\sum_{i<\alpha} \omega^{a_i} r_i + y'\}|\{\omega^{a_i} r_i + \omega^{a_j}\varepsilon\}$ which is x by lemma 5.10(3).

Now suppose $g(y^{o})$ does not contain a minus. Then any upper element $[g(x)]"$ in the canonical representation of $g(x)$ corresponds to a minus in $g(\sum_{i<\alpha} \omega^{a_i} r_i)$, i.e is an upper element in the canonical representation of the latter. Now $g(\sum_{i<\alpha} \omega^{a_i} r_i) = \sum_{i<\alpha} \omega^{a_i} r_i$ $= \{\sum_{i\leqslant\gamma} \omega^{a_i} r_i - \omega^{a_\gamma}\varepsilon\}|\{\sum_{i\leqslant\gamma} \omega^{a_i} r_i + \omega^{a_\gamma}\varepsilon"\}$ where $\gamma < \alpha$ and $\varepsilon', \varepsilon"$ are positive dyadic with numerator 1.

By the inverse cofinality theorem for arbitrary $(g(x))"$ there exists γ and ε such that

$[g(x)]" \geq \sum_{i\leqslant\gamma} \omega^{a_i} r_i + \omega^{a_\gamma}\varepsilon = g[\sum_{i\leqslant\gamma} \omega^{a_i} r_i + \omega^{a_\gamma}\varepsilon]$ by the main inductive hypothesis.

(Note that since ε is dyadic it is guaranteed that $rn\ell[\sum_{i\leqslant\gamma} \omega^{a_i} r_i + \omega^{\gamma}\varepsilon] \leq rn\ell[\sum_{i<\alpha} \omega^{a_i} r_i]$.)

We must now show that $g[\sum_{i\leqslant\gamma} \omega^{a_i} r_i + \omega^{\gamma}\varepsilon]$ is greater than $[g(\sum_{i<\alpha} \omega^{a_i} r_i)][g(y^{o})] = g(x)$.

Of course $g[\sum_{i\leqslant\gamma} \omega^{a_i} r_i + \omega^{a_\gamma}\varepsilon] = \sum_{i\leqslant\gamma} \omega^{a_i} r_i + \omega^{a_\gamma}\varepsilon > \sum_{i<\alpha} \omega^{a_i} r_i$ $= g(\sum_{i<\alpha} \omega^{a_i} r_i)$. Thus by the lexicographical order the only difficulty arises when $\sum_{i<\alpha} \omega^{a_i} r_i$ is an initial segment of $\sum_{i\leq\gamma} \omega^{a_i} r_i + \omega^{a_\gamma}\varepsilon$. In order for this to occur r_γ is necessarily dyadic. $\sum_{i\leq\gamma} \omega^{a_i} r_i + \omega^{a_\gamma}\varepsilon$ is the sequence $\sum_{i\leq\gamma} \omega^{a_i} r_i$ followed by $\omega^{a_\gamma^+}$ pluses and $\omega^{a_\gamma^+} n$ minuses for some integer n. Therefore the contribution of $\sum_{\gamma<i<\alpha} \omega^{a_i} r_i$ to $\sum_{i<\alpha} \omega^{a_i} r_i$ consists only of pluses. Necessarily $\gamma < i \to r_i$ is a positive integer and $\gamma \leq i \to a_{i+1}$ is an initial segment of a_i. The latter follows since $a_{i+1} < a_i$ as, otherwise, a_{i+1} would have a minus not occurring in a_i and, a fortiori, not in any a_j for $j < i$, and would thus contribute a minus to $\sum_{i<\alpha} \omega^{a_i} r_i$. It follows that $\alpha = \gamma+n$ for some integer n. So $\sum_{\gamma<i<\alpha} \omega^{a_i} r_i$ contributes $\sum_{\gamma<i<\alpha} \omega^{a_i^+} r_i$ pluses to $\sum_{i<\alpha} \omega^{a_i} r_i$ where $\gamma \leq i \to a_i^+ > a_{i+1}^+$. Furthermore $g(y^o)$ contains only pluses. Since $r_{\alpha-1}$ is dyadic all minuses in a_α are contained in $a_{\alpha-1}$; hence the number of pluses in $g(y^o)$ is bounded above by a number of the form $\omega^{a_\alpha^+} m$ for some integer m. Since $a_\alpha^+ < a_{\alpha-1}^+$ this finally shows that $g(\sum_{i<\alpha} \omega^{a_i} r_i)g(y^o)$ consists of the sequence $\sum_{i\leq\gamma} \omega^{a_i} r_i$ followed by less than $\omega^{a_\gamma^+}$ pluses. This proves the inequality.

We now have what we need by the cofinality theorem to deduce that $g(x)$ may be expressed in the form $\{\sum_{i<\alpha} \omega^{a_i} r_i + y'\}|\{\sum_{i\leq\gamma} \omega^{a_i} r_i + \omega^{a_\gamma}\varepsilon\}$. Case (3) of lemma 5.10 applies, so the above is a representation of x, thus finally $g(x) = x$.

This completes the induction for the case where $rn\ell(x)$ is a non-limit ordinal. Now suppose $rn\ell(x)$ is a limit ordinal. Then $rn\ell(x) = n\ell(x)$. (In general, if $n\ell(x)$ has the form $\omega a + m$, then $rn\ell(x)$ has the form $\omega a + n$ since the blocks used in the definition of $rn\ell(x)$ are all finite.) Thus we may assume that x has the form $\sum_{i<\alpha} \omega^{a_i} r_i$ for a limit ordinal α and that the result is known to be valid for $\sum_{i<\beta} \omega^{a_i} r_i$ for $\beta < \alpha$. We shall show that the representations of x given by the definition and the canonical representation of $g(x)$ are mutually cofinal which is enough to guarantee that $g(x) = x$. It suffices to consider upper sums since the argument is similar for lower sums.

Let y be an arbitrary upper element of $g(x)$. Then y is an initial segment of x determined by a minus sign contributed by a term $\omega^{a_\beta} r_\beta$ for $\beta < \alpha$. Since α is a limit ordinal, $\beta+1 < \alpha$. Consider $z = \sum_{i \leq \beta+1} \omega^{a_i} r_i + \omega^{a_{\beta+1}} \varepsilon$ which is an upper element of x. By the inductive hypothesis this is $g(z)$ which contains $g(\sum_{i \leq \beta} \omega^{a_i} r_i)$ as an initial segment which in turn contains y followed by a minus as an initial segment. Hence $z < y$ by the lexicographical order.

Now let y be an arbitrary upper element of x. Then y has the form $\sum_{i \leq \beta} \omega^{a_i} r_i + \omega^{a_\beta} \varepsilon$ where we may assume that ε is a dyadic with numerator one.

We claim first that the set of γ for which $\omega^{a_\gamma} r_\gamma$ contributes a minus to $g(x)$ is cofinal in α. Otherwise suppose there exists a γ such that the contribution of $\sum_{\gamma < i < \alpha} \omega^{a_i} r_i$ to $g(\sum_{i < \alpha} \omega^{a_i} r_i)$ consists only of pluses. As in the proof of the case where α is a non-limit ordinal we obtain that a_{i+1} is an initial segment of a_i for $\gamma \leq i$. However, since α is a limit ordinal this already is a contradiction.

It follows that $g(x)$ cannot be an initial segment of y since otherwise all contributions of $\omega^{a_i} r_i$ for $\beta < i$ to $g(x)$ would consist only of pluses which we just noted is impossible.

Hence $g(x)$ is defined at the least ordinal j for which $g(x)$ and y differ. The sign $[g(x)](j)$ is contributed from a term $\omega^{a_\gamma} r_\gamma$. Suppose $(g(x))(j) = +$. By the lexicographical order this would imply that $y < g(\sum_{i \leq \delta} \omega^{a_i} r_i) = (\sum_{i \leq \delta} \omega^{a_i} r_i)$ where $\delta = \max(\beta+1, \gamma)$ which is false. Hence $(g(x))(j) = -$. By what was said earlier there exists $\delta > \gamma$ such that $\omega^{a_\delta} r_\delta$ also contributes a minus to $g(x)$. This determines an upper element z of $g(x)$. $g(z)$ contains $g(\sum_{i \leq \gamma} \omega^{a_i} r_i)$

as an initial segment since $\delta > \gamma$. Hence by the lexicographical order $z = g(z) < y$ which is precisely what we need.

This finally completes the proof that $g(x) = x$ in all cases.

Now that we have the fundamental relation between the sign sequence of a surreal number and its normal form, we can study the surreal numbers in more detail. For this purpose it will be convenient to express theorem 5.11(a) in a form which considers contributions to the sign sequence of ω^a by strings of pluses and minuses. Specifically, if a string begins at the ith place in a then the next string begins at the jth place where j is the least ordinal larger than i such that $a(j) \neq a(i)$.

Corollary 5.1. The sign sequence of ω^a consists of the following juxtaposition. We begin with a plus. For each string of pluses in a we have a string of ω^b pluses where b is the total number of pluses in a up to and including the string. For each string of minuses in a we have a string of $\omega^{b+1}c$ minuses where b is the total number of pluses in a up to the string and c is the number of minuses in the string. a is regarded as beginning with a string of pluses.

Proof. This follows immediately from theorem 5.11(a). For pluses we use the identity $\sum_{\beta<i<\alpha} \omega^{i+1} = \omega^\alpha$ for $\beta < \alpha$. For minuses a_α remains fixed during a string. (No minus contributes a plus!) Finally if a begins with a minus then a may be regarded as beginning with 0 pluses giving rise to $\omega^0 = 1$ plus in ω^α. Thus the last statement is a convenient way of unifying the cases where a begins with a plus and where a begins with a minus. In the former case the first plus is superfluous by absorption, so the statement gives the plus precisely when it should.

To illustrate this we refer back to the second example following the statement of theorem 5.11. We had $a_0 = (-+-++)$. The string of two pluses at the end gives rise to ω^3 pluses since there is a total number of three pluses up to and including that string.

Note finally how strings of pluses and strings of minuses are treated in entirely different ways.

6 LENGTHS AND SUBSYSTEMS WHICH ARE SETS

Up to now we have considered basic material which in some form is contained in [1]. At this point we begin to consider problems that are new and more specialized. In this chapter we are interested in information regarding upper bounds of lengths of surreal numbers obtained by various operations. This will allow us to obtain subclasses of the proper class of surreal numbers which are actually subsets and closed under desirable operations. Our first result is an easy one.

Theorem 6.1. $\ell(a+b) \leq \ell(a) \oplus \ell(b)$.

Proof. We use induction as usual. A typical upper or lower element in the canonical representation of $a+b$ has the form a^o+b or $a+b^o$. Without loss of generality consider a^o+b. By the inductive hypothesis $\ell(a^o+b) \leq \ell(a^o) \oplus \ell(b) < \ell(a) \oplus \ell(b)$. The result follows by theorem 2.3.

This result is sharp. In fact, we already know that in the special case where a and b are ordinals we obtain equality. If a and b are not ordinals then we usually have a proper inequality. For example, $\ell(1) = 1$ but $\ell(\frac{1}{2}) = 2$, hence $\ell(1) < \ell(\frac{1}{2}) + \ell(\frac{1}{2})$. Incidentally, the sign sequence formula of the preceding chapter makes it comparatively routine to study the question of when equality is obtained. We do not pursue this here since we are at present more interested in bounds.

Theorem 6.2. $\ell(ab) \leq 3^{\ell(a) \oplus \ell(b)}$. [We are using ordinary ordinal exponentiation.]

Proof. Using induction as in the proof of theorem 6.1 we must consider elements of the form $a^ob + ab^o - a^ob^o$.

Since $\ell(-a) = \ell(a)$ and since theorem 6.1 extends trivially to arbitrary finite sums we have: $\ell(a^o b + ab^o - a^o b^o) \leq \ell(a^o b) \oplus \ell(ab^o) \oplus \ell(a^o b^o)$. Let $c = \max[\ell(a) \oplus \ell(b^o), \ell(a^o) \oplus \ell(b)]$. Then by the inductive hypothesis $\ell(a^o b) \leq 3^c$ and $\ell(ab^o) \leq 3^c$. Clearly $\ell(a^o) \oplus \ell(b^o) < c$. Hence $\ell(a^o b^o) < 3^c$. Therefore $\ell(a^o b + ab^o - a^o b^o) < 3^c + 3^c + 3^c = 3^{c+1}$. Furthermore $c < \ell(a) \oplus \ell(b)$, hence $c+1 \leq \ell(a) \oplus \ell(b)$. We have shown that an arbitrary element in the canonical representation of ab has length less then $3^{\ell(a) \oplus \ell(b)}$, thus the result follows by theorem 2.3. [Note that 3^c is necessarily a monomial in the expansion of ordinals in terms of powers of ω so that $\oplus$ and $+$ agree with addends of the form 3^c, hence the above computation is justified.]

It is strongly conjectured that $\ell(ab) \leq \ell(a) \times \ell(b)$ in analogy with the case for addition. In fact, it appears as if a routine but extremely messy proof is possible using the sign sequence formula (theorem 5.12). However, I feel that if the conjecture is true there should exist an elegant proof. [In fact, the referee of the original manuscript claims to have a proof. Since, as is pointed out later, theorem 6.2 is adequate for the future, this is not being pursued here.] It is possible to prove theorem 6.1 using the sign sequence formula. In view of the simplicity of our proof this would be silly. On the other hand, such a proof would give us detailed information on the comparison of $\ell(a+b)$ and $\ell(a) \oplus \ell(b)$.

Fortunately theorem 6.2 suffices for important future results in spite of the crudeness of the bound used. For example, to begin with we have the following corollary.

Corollary 6.1. The set of surreal numbers with length less than a fixed ε number is a subring of the class of surreal numbers.

The situation for reciprocals is more complicated. The fact that $\ell(3) = 3$ and $\ell(\frac{1}{3}) = \omega$ makes it clear that a different type of result is needed. Specifically we shall deal with the cardinality of the lengths.

Theorem 6.3. $|\ell(\frac{1}{a})| \leq \aleph_0 |\ell(a)|$ [i.e. unless $\ell(a)$ is a finite ordinal $|\ell(\frac{1}{a})| \leq |\ell(a)|$.]

Proof. We use the construction in chapter 3C. Let $d = \max(|\ell(a)|, \aleph_0)$. The collection of all elements of the form $\langle a_1, a_2, \cdots, a_n \rangle$ used in the above construction has cardinality bounded above by d since all a_i are initial segments of a.

Our main induction will be on $\ell(a)$ and we shall show that $|\ell\langle a_1, a_2, \cdots, a_n \rangle| \leq d$ using a subsidiary induction on n. Consider $\langle a_1, a_2, \cdots, a_{n+1} \rangle$ and let $b = \langle a_1, a_2, \cdots, a_n \rangle$. Then $\langle a_1, a_2, \cdots, a_{n+1} \rangle$ is the unique solution of the equation $(a-a_{i+1})b + a_{i+1}x = 1$. $|\ell(a_{i+1})| \leq d$. $|\ell(b)| \leq d$ by the subsidiary inductive hypothesis and $|\ell(\frac{1}{a_{i+1}})| \leq d$ by the main inductive hypothesis. Hence by theorems 6.1, 6.2 and elementary facts about ordinals $|\ell(x)| \leq d$. (Note that for infinite ordinals α, $|3^\alpha| = |\alpha|$.)

Hence the upper and lower elements in the representation of a consist of at most d numbers each of which has length of at most cardinality d. Hence there is a single ordinal α of cardinality d which is an upper bound to the lengths of all the upper and lower elements. (Such an ordinal is obtained as an ordinal sum of all the lengths since $d^2 = d$.)

Finally $\ell(\frac{1}{a}) \leq \alpha+1$ by theorem 2.3 and certainly $|\alpha+1| = |\alpha| \leq d$.

Theorem 6.3 can also be proved by using the normal form in a manner similar to the proof we shall use for theorem 6.4. I feel that it is worthwhile to have proofs of both kinds, i.e. those that use the normal form and those that don't. Even if there is a known proof of one type for a certain theorem a search for a proof of the other type can lead to certain new insights. It is tempting to make an analogy with synthetic and analytic proofs in geometry, with analytic proofs corresponding to proofs which use the normal form. This makes some sense although analogies always have their limitations.

Theorem 6.4. If the cardinality of the lengths of all the coefficients in a polynomial of odd degree is bounded above by an infinite d, then the polynomial has a root b such that $|\ell(b)| \leq d$.

The result will follow from several lemmas dealing with lengths of elements in normal form. All these lemmas follow from theorem 5.12.

Lemma 6.1. If a is not dyadic then $|\ell(\omega^a)| = |\ell(a)|$.

Proof. Since every term in a gives rise to at least one term in ω^a (in fact at least ω terms), it is clear that $\ell(a) \leq (\omega^a)$ hence certainly $|\ell(a)| \leq |\ell(\omega^a)|$.

Now given any surreal number a, let b be obtained from a by replacing all minuses in a by pluses. [Although formally we can say that $b = \ell(a)$, it is preferable to keep surreal ordinals and ordinals regarded as lengths apart, especially because the algebraic operations are different.]

Since the construction of the sign sequence for ω^a involves counting the number of pluses at various stages, clearly $\ell(\omega^a) \leq \ell(\omega^b)$. However, $|\ell(\omega^b)| = |\ell(b)| = |\ell(a)|$. Hence $|\ell(\omega^a)| \leq |\ell(a)|$. (Recall the fact that $\sum_{\alpha<\beta} \omega^{\alpha+1} = \omega^\beta$.)

Lemma 6.2. $|\ell(\omega^a r)| = |\ell(\omega^a)|$ for any non-zero real r if $a \neq 0$.

Proof. Since ω^a is an initial segment of $\omega^a r$ it follows that $\ell(\omega^a) \leq \ell(\omega^a r)$.

Now $\omega^a r$ is obtained by following ω^a by at most ω strings each of which is obtained from ω^a by ignoring minuses. Hence $\ell(\omega^a r) \leq [\ell(\omega^a)][\omega]$. Therefore $|\ell(\omega^a r)| \leq \aleph_0 |\ell(\omega^a)|$. If $a \neq 0$ then ω^a has infinite length. This completes the proof.

Lemma 6.3. If $\omega^a r$ is one of the terms in the normal form of b then $\ell(\omega^a) \leq \ell(b)$.

Note first that this is not as obvious as it looks. If $\omega^a r$ is the first term then it is an initial segment of b so the result is immediate. In general the difficulty is caused by the fact that because of the ignoring of minuses the contribution of ω^a to the sign sequence of b is not necessarily the sign sequence of ω^a.

Consider the example $-\omega^{-1} + \omega^{-2}$. This consists of a minus followed by $\omega+1$ pluses and ω minuses. ω^{-2} consists of a plus followed by $\omega.2$ minuses. Thus ω^{-2} cannot be a subsequence of $-\omega^{-1}+\omega^{-2}$ but they both have the same length.

Nevertheless the idea of the proof is not too hard. If a

minus is ignored in the term $\omega^{a}r$ there is a corresponding minus which makes a contribution earlier. (There is an exception which is handled separately, namely that of the extra minus ignored.)

Proof. We show that $\omega^{a}r$ is a subsequence of b providing the distinction between pluses and minuses in the sign sequence is ignored. Consider a minus occurring in a which is ignored in the contribution of $\omega^{a}r$ to b. Then the initial segment of a consisting of the sequence up to and including this minus occurs as an initial segment of an a_α for a term $\omega^{a_\alpha}r_\alpha$ preceding the term $\omega^{a}r$, or the term $w^{a}r$ has a predecessor which has the form $w^{a_\alpha}r_\alpha$ with a_α the initial segment up to but not including the minus and with r_α non-dyadic. For each minus such that the first condition holds we choose the first term satisfying the condition. Thus corresponding to the sequence of minuses in $\omega^{a}r$ which are ignored, except possibly for the last one, we obtain a sequence $t_1,t_2\cdots t_\beta$. If $\gamma < \delta$ then $t_\gamma \leq t_\delta$ since the condition for agreement is more stringent for larger γ. It is not necessarily strictly increasing. For each t in the sequence let $D_t = \{\alpha : t_\alpha = t\}$. Then D_t is a partition of all the minuses ignored except possibly the last one. If $s < t$, $\gamma \in D_s$ and $\delta \in D_t$ then $\gamma < \delta$. We now obtain the following subsequence of b by juxtaposition. From each term $\omega^{a_t}r_t$ we extract the contribution to ω^{a_t} of those signs such that the first minus not preceding it corresponds to a term in D_t. By brute force this juxtaposition gives the part of the sign sequence of $\omega^{a}r$ up to and including the contribution of all minuses in a which are ignored except possibly for the qualification mentioned above.

We still must consider the complication caused by minus signs which are ignored although they do not occur previously. This may happen in $\omega^{a_t}r_t$ as well as $\omega^{a}r$. In this case we use the rest of the contribution of the preceding term. The lack of any contribution from the ignored minus sign is precisely made up by the contribution from the preceding non-dyadic coefficient. Recall that if c has α pluses and if d is c followed by a minus then the minus contributes $\omega^{\alpha+1}$ signs to ω^{d}. This is the same as $\omega^{\alpha}\omega$ which is the number of signs contributed by a non-dyadic r to $\omega^{c}r$ since r has ω signs, each contributing ω^{α} signs.

Lemma 6.4. Let $a = \sum_{\alpha<\beta}^{a} \omega^{a_\alpha} r_\alpha$. Then $|\beta| \leq |\ell.u.b.\ [\ell(a_\alpha)\omega]|$.

Remark. It turns out that this lemma is not needed for the proof of the theorem. However, I feel that it is of interest in itself and fits in naturally in our present discussion of lengths. A similar remark applies to lemma 6.5. What is really needed is lemma 6.5a which strengthens both the hypothesis and conclusion of lemma 6.5. Lemma 6.5 is of interest because of the surprising fact that the bound is sharp in spite of the lack of any hypothesis on the length of β.

Proof. If $\ell(a_\alpha)$ is finite for all α then there can be at most ω terms so we need only consider the case where at least one a_α has infinite length.

Let $d = |\ell.u.b.\ \ell(a_\alpha)|$. Then all a_α terminate at d or earlier. By a simple cardinality argument there are at most 2^d such sequences. This is not good enough for our purpose, and in fact we shall show that this can be strengthened using the fact that the a's are well-ordered.

In fact, assume that there exists a well-ordered decreasing sequence of surreal numbers (a_α) containing $d^+ > d$ members where d^+ is the successor of d. We show that this leads to a contradiction.

First consider the sequence $a_\alpha(0)$. Since this can take on only the values plus, minus, and 0 by the lexicographical order there exists an a_α such that $a_\alpha(0)$ is fixed for $\alpha > \alpha_0$. [E.g. if $a_\alpha(0)$ is minus for any α it must be minus for all larger α.] If the fixed value is 0 we have an immediate contradiction since this says in particular that $a_{\alpha_0} = a_{\alpha_0+1} = 0$. (Recall that 0 stands for "undefined.")

We now construct a monotonic increasing sequence (α_i) defined for all $i \leq \ell.u.b.\ \ell(a_\alpha)$ satisfying $a_\alpha(i) = a_{\alpha_i}(i)$ for all $\alpha > \alpha_i$.

Suppose α_i is defined for $i \leq j$. Since α_i is monotonic increasing we know that $a_\alpha(i) = a_{\alpha_j}(i)$ for all $\alpha > \alpha_j$ and all $i \leq j$. Consider the subsequence $a_\alpha(j+1)$ where $\alpha \geq \alpha_j$. By the lexicographical order this is monotonic decreasing. Hence the same argument as in the case $a_\alpha(0)$ shows that there exists an α_{j+1} such that $a(j+1) = a_{\alpha_{j+1}}(j+1)$ for all $\alpha \geq \alpha_{j+1}$. Note that the subsequence considered still has d^+ members.

Now suppose j is a limit ordinal and α_j is defined for all $i < j$. Since $|j| \leq d$ the set of all α_i for $i < j$ is bounded above by a certain β. By the same argument as before considering the subsequence of a_α for $\alpha \geq \beta$ we can find a suitable α_j.

Finally consider the case $i = \ell.u.b.\ \ell(a_\alpha)$. The construction gives us an α_i such that for all $\alpha > \alpha_i$ we have $a_\alpha(j) = a_{\alpha_i}(j)$ for $j \leq i$. However, $a_\alpha(j)$ is undefined for $j > i$ (even for $j=i$). Hence $a_\alpha = a_{\alpha_i}$ for $\alpha > \alpha_i$. Since all a_α are distinct this is a contradiction.

Remark: Note that the theorem refers to the cardinality of the $\ell.u.b.$ of a set of ordinals. This is not the same as the $\ell.u.b$ of the cardinalities of the set of ordinals, so caution is required in the statement of the theorem. To see the distinction it suffices to consider the case where the set consists of all the countable ordinals. Following this through if we let $a = \sum_{\alpha<\omega_1} \omega^{a_\alpha}$ where a_α consists of a plus followed by α minuses then we see that the conclusion of lemma 6.4 cannot be expressed as $|\beta| \leq \ell.u.b.\ |\ell(a_\alpha)|$ since the right-hand side is $\aleph_0$.

Lemma 6.5. If $a = \sum_{\alpha<\beta} \omega^{a_\alpha} r_\alpha$ then $|\ell(a)| \leq |\ell.u.b.\ \ell(a_\alpha), \omega|$.

Proof. By lemmas 6.1 and 6.2 we know that $|\ell(\omega^{a_\alpha} r_\alpha)| \leq |\ell(a_\alpha)|\aleph_0 \leq d$ where $d = |\ell.u.b\ \ell(a_\alpha), \omega|$. Hence the contribution of $\omega^{a_\alpha} r_\alpha$ to the sign sequence of a has at most cardinality d since signs may be ignored but no extra signs added. By lemma 6.4 there are at most d terms in the expansion of a. Hence there are at most $d^2 = d$ signs in the sign sequence for a.

The following corollary is immediate from lemmas 6.3 and 6.5.

Corollary 6.2 If $a = \sum_{\alpha<\beta} \omega^{a_\alpha} r_\alpha$ where not all a_α are dyadic then $|\ell(a)| = |\ell.u.b\ \ell(a_\alpha)|$.

Lemma 6.5a. If $a = \sum_{\alpha<\beta} \omega^{a_\alpha} r_\alpha$ and $\ell.u.b.\ (|\ell(\beta)|, (\ell(a_\alpha)|, \aleph_0) \leq d$ then $|\ell(a)| \leq d$.

Proof. This is the same as the last part of the proof of lemma 6.5. Note that the bound here is slightly sharper than the one in lemma 6.5. This bound we actually have to "pay for" although the early bound came free thanks to lemma 6.4.

Lemma 6.6. If $\alpha_1, \alpha_2 \ldots \alpha_n$ are arbitrary surreal numbers and $r_1, r_2 \ldots r_n$ are rational then $|\ell(\sum r_i\alpha_i)| \leq |\max \ell(\alpha_i)|\aleph_0$.

Proof. We know that for any real r $\ell(r) \leq \omega$ so $|\ell(r_i)| \leq \aleph_0$ for all i. The lemma then follows from theorems 6.1 and 6.2. (Of course, all that is used is a weakened form of these theorems which refer only to cardinalities of lengths.)

We now have all we need to prove theorem 6.4. This may seem strange since none of the lemmas have anything to do with polynomials of odd degree! In fact, the proof of the theorem will not make direct reference to such polynomials. The aspect of theorem 5.10 which is crucial is that the exponents are rational linear combinations of the given exponents, so that the same proof works for reciprocals.

Proof of Theorem 6.4. This follows easily from the lemmas by a kind of back and forth argument.

Let α be an ordinal of infinite cardinal d which is an upper bound to the lengths of all the coefficients. By lemmas 6.3 and 6.1 (used in that order) α is also an upper bound to the lengths of the exponents occurring in the normal forms of the coefficients. Since $n\ell(a) \leq \ell(a)$ for any surreal number, α is also an upper bound to the normal lengths of all the coefficients. By lemma 6.6 d is an upper bound to the cardinalities of the lengths of the exponents occurring in the normal form of the constructed root. d is also an upper bound to the cardinality of the normal length of the constructed root. (By elementary cardinal arithmetic, the cardinality of the set of all finite rational linear combinations of an infinite set S has the same cardinality as S.) By lemma 6.5a the cardinality of the length of the constructed root is bounded above by d.

Remark. A proof using lemma 6.5 is also possible although our approach seems simpler. For such a proof we need a strengthened form of lemma 6.6

which makes fuller use of theorems 6.1 and 6.2. This will enable us to obtain an ordinal upper bound of cardinality d to the lengths of the exponents occurring in the normal form of the constructed root. In fact the least ε number larger than α works.

Finally, as a culmination of the results of this chapter we have shown that the subset of surreal numbers a such that $|\ell(a)| \leq d$ for any fixed infinite cardinal d is a real closed field. Since all operations concerned depend on only finitely many elements the condition $\ell(a) \leq d$ may be replaced by $\ell(a) < d$. [The latter formulation gives more fields.] These are all "honest" fields since their carriers are sets.

Not all subfields have the above form. In fact, the two most well-known fields found in nature, the rationals and reals, both consist of all surreals of finite length together with some but not all surreals of length ω.

The field of all surreals of countable length should be a worthwhile object for further study.

7 SUMS AS SUBSHUFFLES, UNSOLVED PROBLEMS

This chapter is still at a pioneering level. I have a strong feeling that there is a right way of looking at the subject which when discovered will greatly enrich the theory of surreal numbers.

A sequence $\langle c_0, c_1, \cdots c_\gamma\rangle$ is a shuffle of the sequences $\langle a_0, a_1, \cdots a_\alpha\rangle$ and $\langle b_0, b_1 \cdots b_\beta\rangle$ if there exist strictly increasing sequences $\langle i_1, i_2, \cdots i_\alpha\rangle$ and $\langle j_1, j_2 \cdots j_\beta\rangle$ such that $(\forall k \leq \alpha)(a_k = c_{i_k}) \wedge (\forall k \leq \beta)(b_k = c_{j_k})$ and such that every ordinal not larger than γ is one of the i's or j's but not both. A subshuffle of two sequences is a subsequence of a shuffle of the two given sequences.

This definition is consistent with the intuitive meaning of a shuffle. For ordinals it is known that $\alpha + \beta$, where $+$ refers to surreal addition, is the largest ordinal which can be obtained as a shuffle of α and β. For example, if $\alpha = \omega+5$ and $\beta = \omega+3$ then the "trivial" shuffles $\alpha+\beta$ and $\beta+\alpha$ give $\omega\cdot 2+3$ and $\omega\cdot 2+5$ respectively. However, the shuffle $\omega+\omega+5+3$ (i.e. we first take the ω from α then the ω from β and then take the rest of α followed by the rest of β) gives $\omega\cdot 2+8$ which is $\alpha+\beta$.

This chapter is devoted to the proof of the following theorem which strengthens theorem 6.1.

Theorem 7.1. For any surreal numbers a and b, $a+b$ is a subshuffle of a and b.

Proof. (a) We do this first in the case where $\ell(a)$ and $\ell(b)$ are finite, using induction. (This seems easier than the use of arithmetic of dyadic fractions as discussed at the end of chapter 4B, where the strange nature of the carrying causes complications.) For the purpose of the proof it is convenient to think of a subshuffle of a and b informally as a sequence obtained by moving along, selecting a sign from either a or b with the restriction that the signs from a and b must be taken in the order they occur although signs may be omitted.

Now $a+b = \{a'+b, a+b'\} | \{a''+b, a+b''\}$. Suppose a+b ends with a plus (a+b is a finite sequence). A similar argument will apply if it ends in a minus.

Let a+b = d+. By the inverse cofinality theorem there exists an a' such that $a'+b \geq d$ or there exists a b' such that $a+b' \geq d$. Since both cases are similar it suffices to assume that $a'+b \geq d$.

Since $a+b > a'+b \geq d$, a'+b must have d as an initial segment. (This includes the possibility d = a'+b.) By the inductive hypothesis a'+b and hence a fortiori d is a subshuffle of a' and b. Now a' is by definition a subsequence of a obtained by stopping at a plus, i.e. this is a plus in a occurring after all the signs in a'. Thus d+ can be obtained as a subshuffle of a and b by using the representation of d as a subshuffle of a' and b followed by that plus.

As a technical detail for the future we need the following: If $a > 0$, $b < 0$, and $a+b > 0$ then the subshuffle for a+b may be chosen so that the first plus comes from a. This is not a trivial requirement. For example, let a = (+-), b = (-++--) and c = (++--+). Then c can be expressed as a subshuffle of a and b by beginning with the segment (++--) of b. However, if one tries to begin with the plus in a one gets stuck at the last plus in c. Of course $c \neq a+b$ since it is immediate that $c > 1$ and that $a+b < 1$.

By hypothesis a begins with a plus, b with a minus, and a+b with a plus. In the inductive hypothesis we consider terms such as a^o, a^o+b, etc. (Recall that a^o is either of the form a' or a''.) First suppose that $a^o > 0$, and $a^o+b > 0$. Then the triple (a^o, b, a^o+b) satisfies the technical hypothesis. Since the subshuffle used for a+b obtained in the proof begins with the subshuffle used for a^o+b it follows that the first plus in a+b is taken from a.

To complete the proof of the technical detail we must examine the cases where terms such as a^o and a^o+b do not satisfy the hypothesis. Recall in the proof that a^o+b has d as an initial segment where

$a+b = d$ followed by a plus [or d followed by a minus if a^o has the form a'']. Hence the only cases where terms such as a^o and a^o+b fail to satisfy the technical hypothesis occur when terms such as a^o, b^o, or d are 0.

If $b^o = 0$, it certainly contains no pluses so the first plus in $a+b^o$ is necessarily taken from a (i.e. no inductive hypothesis is needed).

If $d = 0$ then $a+b = (+)$ [$a+b > 0$]. The plus can certainly be taken from a since a is positive and thus contains at least one plus, for example the first sign.

If $a^o = 0$ then $a^o+b = b < 0$. Since $a+b > 0$ and both $a+b$ and a^o+b begin with d it follows that $d = 0$; thus we are back in the previous case.

Of course a similar result applies if all signs are reversed.

(b) It is now easy to see that the result is valid for all real a and b. Since the finite case is the same as the dyadic case it suffices to assume that a is not dyadic. This case is now trivial. Since a has both pluses and minuses arbitrary far out and $a+b$ has length ω we can obtain $a+b$ from a alone; in fact no matter what we do we can't possibly get stuck because of the availability of signs from a. In particular, the technical detail referred to earlier can be satisfied.

(c) We now prove this for numbers of the form $\omega^a r$ and $\omega^a s$. Recall that the sign sequence for $\omega^a r$ is given by a contribution from ω^a followed by the contribution from r which resembles r itself except that the first sign is ignored and each other sign from r is repeated ω^α times where α is the number of pluses in a. If the first sign were not ignored in the rule the proof would follow trivially from (b). The subshuffle for $\omega^a(r+s)$ could be simply obtained by selecting the contribution of ω^a from a suitable one of $\omega^a r$ or $\omega^a s$ and each block of signs in the contribution of $r+s$ from the corresponding place where the corresponding sign in $r+s$ is obtained. This is clearly a subshuffle if the contribution of ω^a is selected from a term whose coef-coefficient has the same sign as $r+s$. (At least one of r and s has the same sign as $r+s$. Since the result is vacuously true when $r+s = 0$ we can ignore that case.)

We have to adjust the construction to take account of the fact that the first signs of the coefficients are ignored in the sign sequence rule. (Of course, we do not mean that it is totally ignored. The first sign determines whether $\omega^a r$ is positive or negative and thus affects the contribution of ω^a.)

Let x,y, and z be sequences corresponding to $\omega^a r$, $\omega^a s$, and $\omega^a(r+s)$ in which the first signs are not ignored so that by the above z is a subshuffle of x and y. We show how this can be used to obtain $\omega^a(r+s)$ as a subshuffle of $\omega^a r$ and $\omega^a s$. We break this up into cases.

Suppose r and s are positive. Then so is r+s. The contribution of ω^a consists of ω^α pluses with possibly some minuses interspersed among them. We need less but there is also less to choose from. In expressing a sequence as a subshuffle of other sequences no harm can ever be done by selecting at any time the earliest place where a sign occurs in the sequence (since any continuation which was legal before is a fortiori still legal). We may therefore assume that in the representation of r+s as a subshuffle of r and s the first plus is taken either as the first plus in r or the first plus in s. Suppose without loss of generality that it is the first plus in r. The first plus in s may also have been used in the representation of r+s. (If not our work is easier.) If so, then s could not have been used for earlier terms in the representation by the definition of a shuffle. We now obtain the representation of $\omega^a(r+s)$ as a subshuffle as follows. The contribution of ω^a is taken from the contribution of ω^a in $\omega^a r$. The section in z and x which is missing in $\omega^a(r+s)$ and $\omega^a r$ respectively is ignored. The only possible difficulty remaining occurs if the representation of z makes use of the segment of y which is not contained in $\omega^a s$. i.e. the first plus in s is used in the representation of r+s. We then still can obtain a subshuffle by using the ω^α pluses contributed by ω^a in $\omega^a s$ since they have not yet been used, and by the above, the ordering requirement is still maintained.

If r and s are negative the result follows by sign reversal.

Now suppose r is positive and s is negative. Also suppose without loss of generality that r+s is positive. The construction is similar to the previous one thanks to the technical condition referred to in part (a). We may thus assume that the first plus in r+s is taken as the first plus in r. The only possible difficulty now occurs if the minus which s begins with is used in the representation of r+s. Just as in the previous case we can now use the ω^{α} minuses which ω^{a} contributes to $\omega^{a}s$. (Since s is negative these ω^{α} minuses are present by sign reversal because of the ω^{α} pluses in ω^{a}.)

(d) We now prove the result in general. If we allow the use of zero coefficients, we may assume that $x = \sum_{i<\alpha} \omega^{a_i} r_i$, $y = \sum_{i<\alpha} \omega^{a_i} s_i$ and $x+y = \sum_{i<\alpha} \omega^{a_i}(r_i+s_i)$. Now if no minuses were ignored in the sign sequence formulas the result would be trivial. Juxtaposition of all the subshuffles as we run through all i would lead to a subshuffle of x and y. Furthermore, minuses "missing" in $x+y$ cause no trouble; in fact, this makes things easier. The difficulties occur only when minuses in x or y are "missing." As in the proof of part (c) we must show that other sources for these minuses exist so that $x+y$ can still be obtained as a subshuffle. (It is ironic that the main cases of difficulty occur when $r_i+s_i = 0$ for some i! Since the corresponding term in $x+y$ vanishes, the subshuffle requirement appears to be vacuous. However, the issue of ignored minuses arises in later terms when the above equality holds.)

We shall obtain $x+y$ as a subshuffle of x and y by induction on α. We assume that $\sum_{i<\beta} \omega^{a_i}(r_i+s_i)$ has been obtained as a subshuffle of $\sum_{i<\beta} \omega^{a_i} r_i$ and $\sum_{i<\beta} \omega^{a_i} s_i$ and show that $\sum_{i\leq\beta} \omega^{a_i}(r_i+s_i)$ can be obtained as a subshuffle of $\sum_{i\leq\beta} \omega^{a_i} r_i$ and $\sum_{i\leq\beta} \omega^{a_i} s_i$ as a continuation of the earlier subshuffle. (The last part is crucial for reaching limit ordinals. Subshuffles for $i < \gamma$ for all $\gamma < \beta$ do not lead in an obvious way to a subshuffle for $i < \beta$ unless they satisfy the continuation property.)

Suppose that there is a minus in β which is not ignored in the contribution of ω^{a_β} to $x+y$ but is ignored in the contribution of ω^{a_β} to either x or y, say x. Otherwise, as noted before, the continuation of the subshuffle is trivial. We now use an argument which is similar to the one used in the proof of lemma 6.3. Suppose that the minus is ignored in the contribution to x because it occurs in an exponent a_γ for $\gamma < \beta$ in the normal form of x. The other possibility for ignoring minuses will be considered later. Then a_γ does not occur as an exponent in the normal form of $x+y$. Furthermore every exponent between a_γ and a_β also contains the minus so it cannot occur in the normal form of $x+y$ otherwise the minus would be ignored in the contribution of ω^{a_β}. Thus $\gamma \leq \delta < \beta \Rightarrow r_\delta + s_\delta = 0$, i.e. $r_\delta = -s_\delta$.

We can now use the same type of construction used in the proof of lemma 6.3 The contributions of ω^{a_δ} to x and y now give us a supply of pluses and minuses which may be used to obtain the subshuffle needed. As in the proof of lemma 6.3 we first consider the case where all minuses considered do take part in the contribution of ω^{a_γ} for for the least γ in which they occur. Only one little detail is needed to supplement the argument in lemma 6.3. The contribution of the concerned minuses in a_β to $x+y$ depend upon the sign of $r_\beta + s_\beta$. However, since a_δ and b_δ have opposite signs, we have available whichever sign we need.

We now consider a minus not ignored in the contribution of ω^β to $x+y$ but ignored in the contribution to x and not occurring in an exponent $\gamma < \beta$ in the normal form of x. We know that x has an immediately previous term with non-dyadic coefficient. The exponent of that term may occur in $x+y$ but if so it must also be an immediately previous term with a dyadic coefficient (because of the rule for ignoring minuses). Whether or not the exponent occurs in the normal form of $x+y$ it must be the exponent of the previous term of y and have a non-dyadic coefficient. This is the situation we now have. The normal forms of x and y consist of terms $\omega^{a_{\beta-1}} r_{\beta-1}$ and $\omega^{a_{\beta-1}} s_{\beta-1}$ where $r_{\beta-1}$ and $s_{\beta-1}$ are non-dyadic and $r_{\beta-1} + s_{\beta-1}$ is dyadic (in particular it may be 0). $a_{\beta-1}$ may or may not occur as an exponent in the normal form of $x+y$. In any case the inductively defined subshuffle makes use of only a finite number of pluses and minuses in the contributions of $r_{\beta-1}$ and $s_{\beta-1}$ to x and y respectively. (If one is picayune, in order for the proof to be formally correct, this aspect of the construction should be included in the inductive hypothesis. However, an excess of formalism would complicate the exposition unnecessarily.) Recall that a non-dyadic real has pluses and minuses arbitrarily far out. Thus there is enough left over in the contributions of $r_{\beta-1}$ and $s_{\beta-1}$ to the sign sequence of x and y respectively to continue the subshuffle of $x+y$ to include the contribution of the minus. (The number of signs needed was calculated in the proof of lemma 6.3.)

There is still the problem of a similar situation occurring for some $\gamma < \beta$. This can be handled as before by letting the $(\gamma-1)$st term play the role previously played by the $(\beta-1)$st term.

This finally completes the proof of the theorem.

By comparing the proofs of theorem 6.1 and 7.1 we see the contrast between a proof using the normal form and one which does not. I referred earlier to the analogy between this and the contrast between synthetic and analytic proofs in geometry. Very often there is a choice between a lengthy tedious but routine analytic proof and a quick synthetic proof using an appropriate theorem. We see the contrast also in our work on surreal numbers. The use of the normal form gives us a proof which is somewhat routine but messy because of the quirks in the sign sequence formula. However, even if there is a shorter proof which does not use the normal form, the proof using the normal form still may have value because of its constructive nature.

I conjecture that a result similar to theorem 7.1 is valid for multiplication. This would involve orderings obtained from cartesian products of sequences of signs. However, this is far beyond our present pioneering level. Further insights on dealing with sign sequences are needed before one can tackle such problems the right way.

8 NUMBER THEORY

A BASIC RESULTS

This chapter overlaps chapter five in [1] to some extent. We study a subring of the class of surreal numbers for which there are results analogous to those in number theory. By the results of chapter six the theory is essentially unchanged if one restricts oneself to certain suitable subsets of the class of surreal numbers so that we deal with "honest" rings.

Definition. A surreal number a is an integer if the exponents in the normal form of a are all non-negative, and if a zero exponent occurs then the real coefficient is an ordinary integer.

For example, $\omega^{2\frac{1}{3}}+7$ is an integer (the $\frac{1}{3}$ does not prevent this) whereas $\omega^{2}+\frac{1}{4}$ is not. Also, $\omega^{2} + \omega^{-3}7$ is not an integer.

This definition, which is equivalent to the one used in [1], may seem artificial at first. There certainly exists a wide choice of other subrings. However, this definition leads to desirable theorems. Moreover the existence of equivalent definitions, one of which can be expressed simply in terms of the sign sequence and the other in terms of the relation $a = F|G$, suggest on philosophical grounds that our definition leads to a "natural" system.

Theorem 8.1. The following are equivalent

(1) a is an integer.

(2) There does not exist an ordinal α such that $a(\alpha)$ and $a(\alpha+1)$ have opposite signs.

(3) $a = \{a-1\}|\{a+1\}$.

Proof (1) ⇒ (2). This is immediate from the sign sequence formula. If $b > 0$, ω^b begins with an infinite number of pluses, and in $\omega^b r$ all signs are repeated infinitely often. Fortunately, the ignoring of minuses in sums does not cause any extra complications. For $b = 0$ we know that an integer corresponds to a finite sequence of signs which are all alike.

Not (1) ⇒ not (2). For $b = 0$, we know that if r is not an integer, then $r = \omega^b r$ has length at most ω and there is a change in sign somewhere, i.e. we have not (2). If $b < 0$, then ω^b begins with a plus followed by a minus. This almost proves not (2). What remains is to consider the possibility that the minus may be ignored. Let b be the first negative exponent occurring in the normal form of a. If the first minus sign in b is ignored then there is a previous term with exponent zero and non-dyadic coefficient so again we have not (2).
(1) ⇒ (3). First, assume that all exponents are positive. Then the result follows immediately by cofinality, where we use the definition if there is no last term in the normal form and lemma 5.3 if there is a last term. If a zero exponent does occur then it necessarily occurs in the last term with an integer coefficient. The result then follows by cofinality from lemma 5.10. (Actually, since 0 has no proper initial segments we need only a simpler version of the lemma.)
Not (1) ⇒ not (3). Let the normal form consist of terms with negative exponent and express a as $b+c$ where c is the tail in the normal form consisting of all terms with negative exponent. Then by the lexicographical order $a-1 < b < a+1$. Clearly $\ell(a) > \ell(b)$ by the sign sequence formula (in fact, b is a proper initial segment of a), so $a \neq \{a-1\}|\{a+1\}$.

Now let us consider the other possibility, i.e a has the form $\sum_{i<\alpha} \omega^{b_i} r_i + r_\alpha$ where $b_i > 0$ and r_α is not an integer. Then if r_α is replaced by either the largest integer less than r_α or the smallest integer greater than r_α we obtain a number c satisfying $a-1 < c < a+1$. One of these necessarily has smaller length than a, so again $a \neq \{a-1\}|\{a+1\}$.

Each of the equivalent definitions has its own intrinsic interest. (2) requires nothing more than the definition of surreal numbers and thus can even be used in the beginning of chapter two. (3)

is used by Conway in [1]. This is consistent with his style of making the relation $a = F|G$ fundamental. Our definition (1) is also mentioned by Conway. The latter definition appears to be most useful for proofs. In fact, many of the arguments depend only on the generalized group ring structure given by the normal form and make no essential use of the fact that the exponents are surreal numbers. Thus by taking theorems 5.7 and 5.8 as definitions for a set consisting of generalized power series we can obtain a theory which is independent of the theory of surreal numbers although the results will be valid for surreal numbers as one special case. One convenient hypothesis for the exponents is that they form a divisible abelian group.

The following result is immediate by theorems 5.7 and 5.8.

<u>Theorem 8.2</u>. The integers form a subring of the surreal numbers.

<u>Remark</u>. This can also be proved easily using definition (3).

<u>Theorem 8.3</u>. Every surreal number is a quotient of two integers.

<u>Proof</u>. Let $a = \sum_{i<\alpha} \omega^{a_i} r_i$.

Suppose $c = \phi|\{0,a_i\}$. Then $c < 0$ and $(\forall i)(c < a_i)$. Then ω^{-c} is an integer. Now $\omega^{-c}a = \sum_{i<\alpha} \omega^{a_i - c} r_i$. Since $a_i - c > 0$ for all i, $\omega^{-c}a$ is also an integer. Of course $a = \dfrac{\omega^{-c}a}{\omega^{-c}}$.

Let $a = \sum_{i<\alpha} \omega^{c_i} r_i$. For convenience, express this as $y+r+z$, where y consists of all terms with positive exponent, z consists of all terms with negative exponent, and r is a real number which may be 0.

Unless r is an integer and z is negative let $p = y+n$ where n is the largest integer in r.

If r is an integer and z is negative let $p = y+r-1$.

It is clear from the lexicographical order that p is the largest integer not greater than a. It is also clear that p is the unique integer satisfying $p \leq a$ and $a-p < 1$. Incidentally, this remark on the spacing of the integers among the surreal numbers is further justification for our definition of "integer."

Theorem 8.4. The division algorithm holds, i.e. if a and b are positive integers there exist unique integers q and r such that $a = bq+r$ and $0 \leq r < b$.

Proof. Let q be the largest integer not greater than $\frac{a}{b}$. Then $\frac{a}{b} \geq q$ and $\frac{a}{b} - q < 1$. If $r = \frac{a}{b} - q$ it is immediate that $0 \leq r < b$.

Conversely, the conditions $a = bq+r$ and $0 \leq r < b$ imply that $\frac{a}{b} \geq q$ and $q - \frac{a}{b} < 1$. We have already noted that this characterizes q uniquely.

Fortunately, for finite, i.e. ordinary, integers divisibility in the surreal sense is equivalent to ordinary divisibility. However, strange things do happen soon so that our system differs in important ways from ordinary number theory. For example, ω is divisible by every finite integer since $\frac{\omega}{n}$ is an integer for all finite n by our definition. In fact, more generally every integer whose normal form does not contain a term with zero exponent is divisible by every finite integer.

B PARTIAL RESULTS AND UNSOLVED PROBLEMS

We first note that the ring of integers is not Noetherian. In fact, the principal ideals (ω), $(\omega^{\frac{1}{2}})$, $(\omega^{\frac{1}{3}})$ $\cdots$ $(\omega^{\frac{1}{n}})$ form a strictly ascending chain of ideals. Any reader who is dissatisfied with using concepts such as Noetherian for rings which are proper classes should reread the remarks in the first paragraph of this chapter.

Slightly less obvious is the following.

Theorem 8.5. There exists an ideal with two generators which is not principal.

Proof. In fact, let $I = (\omega, \omega\sqrt{2})$. We show by contradiction that I is not a principal ideal. Let $(\omega, \omega\sqrt{2}) = (a)$. If $d < 1$ then ω^d divides both ω and $\omega\sqrt{2}$; hence it divides a. Hence no exponents occurring in the normal form of d can be less than 1.

Now a divides ω, i.e. $\frac{\omega}{a}$ is an integer.

Let $a = \sum_{i<\alpha} \omega^{b_i} r_i$. If we write a in the form $\omega^{b_0} r_0 [1+\sum \omega^{c_i} s_i]$ it is clear that the normal form of $\frac{1}{a}$ begins with $\omega^{-b_0} \frac{1}{r_0}$ (as in dealings with ordinary power series). Therefore the normal form of $\frac{\omega}{a}$ begins with $\omega^{1-b_0} \frac{1}{r_0}$. Since $\frac{\omega}{a}$ is an integer $b_0 \leq 1$. Since we also know that $b_0 \geq 1$ we have $b_0 = 1$. Furthermore it follows that $\frac{\omega}{a}$ can't contain any further terms so $\frac{\omega}{a} = \frac{1}{r_0}$ where r_0 is the reciprocal of an integer, i.e. a has the form $\frac{\omega}{n}$ for some integer n. But then a does not divide $\omega\sqrt{2}$ so we have a contradiction.

We are interested in the classification of primes. By our earlier discussion for ordinary integers, a number is a prime in our system if and only if it is an ordinary prime. It is an open problem whether other primes exist. However, we have partial results which tell us that various surreal numbers are composite so that the search for primes can be narrowed significantly.

For example, let $a = \sum_{i<\alpha} \omega^{b_i} r_i$ be a surreal integer. If $b_i > 0$ for all i then we can let $c = \{0\}|\{b_i\}$. Clearly ω^c divides a so that a is not a prime. Hence the normal form of all primes necessarily end with a term whose exponent is 0. Furthermore, n divides $\sum_{i<\alpha} \omega^{b_i} r_i + n$ so that the last term in the normal form of a prime is necessarily ± 1.

Consider next ordinary polynomials in ω. Every polynomial of degree larger than two factors since any factorization of polynomials obviously leads to a factorization when the indeterminate x is replaced by ω. Furthermore, any polynomial may be regarded as a polynomial of of degree larger than two in $\omega^{\frac{1}{3}}$; hence any polynomial factors, e.g.
$\omega+1 = (\omega^{\frac{1}{3}}+1)((\omega^{\frac{2}{3}}-\omega^{\frac{1}{3}}+1)$.

The same idea can be extended to any finite sum where all the exponents are rational multiples of a fixed surreal a.

The following result, which is a big help in narrowing down our search for primes, is being honoured by being designated as a theorem.

<u>Theorem 8.6</u>. Let $a = \sum_{i<\alpha} \omega^{b_i} r_i$ be an integer. Suppose that there exists

j such that $b_0 >> b_j$ and $b_j \neq 0$. Then a is composite.

Proof. Let k be the least j such that $b_0 >> b_j$. Then we may write $a = x+y$ where $x = \sum_{i<k} \omega^{b_i} r_i$ and where y begins with $\omega^{b_k} r_k$.

Now all exponents in y^{-1} are rational linear combinations of the exponents b_i. Since the b's are decreasing and non-negative, every exponent b_i satisfies $b_0 >> b_i$. Hence all the exponents in y^{-1} have lower order of magnitude than b_0. Furthermore, by choice of k every exponent in x has the same magnitude as b_0. This shows that all exponents in xy^{-1} are positive. (In fact, they have the same order of magnitude as b_0.) Hence xy^{-1} is an integer.

By theorem 8.6 we can narrow our search for primes to those integers all of whose exponents with the exception of the zero at the end have the same order of magnitude, i.e. are finite non-infinitesimal multiples of a fixed surreal a. Furthermore, since a factorization of a number of the form $\sum_{i<\alpha} \omega^{ab_i} r_i$ automatically leads to a factorization of $\sum_{i<\alpha} \omega^{b_i} r_i$, we may just as well limit ourselves to finite non-infinitesimal exponents.

We now mention a device which permits us to regard the exponents as ordinary real numbers but with the price of being stuck with a larger coefficient field.

The class of all surreal numbers in which the exponents occurring in their normal form are all finite forms a subfield since the set of all finite numbers is certainly closed with respect to finite linear rational combinations. Similarly, the class of all surreal numbers in which the exponents occurring in their normal form are all infinitesimal forms a subfield.

Now consider a surreal number a of the form $\sum_{i<\alpha} \omega^{b_i} r_i$ where b_i is finite for all i. It is well known and easy to prove that every finite number a can be expressed uniquely in the form $r+\varepsilon$ where r is real and ε is infinitesimal. (In the language of nonstandard analysis r is called the standard part of a.) We now group together all terms which have the same standard part. Specifically, corresponding to every real number r we obtain an expression of the form $\sum_{i<\beta} \omega^{c_i} s_i$

where c_i) is the subsequence of (b_i) consisting of all b_i whose real part is r. [For some r the set of c_i's may be empty; in fact, because of the well ordering it is not too hard to show that the set is empty except for countably many reals.] This can be expressed as $\omega^r(\sum_{i<\alpha} \omega^{\varepsilon_i} s_i)$ where ε_i is a decreasing sequence of infinitesimals.

As r varies we may thus regard the whole sum as a generalized series with real exponents with coefficients themselves power series with infinitesimal exponents. (Since the set of a's is well-ordered, so is the set of r's.) It is clear from the nature of formal multiplication of series that the latter point of view leads to a system isomorphic to the original.

Thus in a search for primes it is natural to investigate surreal numbers for which the exponents occurring in their normal forms are real numbers.

Among the simplest looking numbers for which the problem reamains open is $\omega\sqrt{2}+\omega+1$.

Open Problem. Is $\omega\sqrt{2}+\omega+1$ a prime?

We suspect that no matter whether the answer turns out to be yes or no, the proof will make no essential use of the number $\sqrt{2}$ but in fact will use only the fact that $\sqrt{2}$ is irrational. Moreover, the proof would probably be extendable to more general polynomials. The following unsolved problem is found in [1].

Open Problem. Is $\omega+\omega^{\frac{1}{2}}+\omega^{\frac{1}{3}} \cdots \omega^{\frac{1}{\pi}} \cdots +1$ a prime?

(Do not let an earlier remark confuse you. We had stated that if all the exponents in a polynomial are rational multiples of a fixed number then the polynomial is composite. The discussion there concerned itself with polynomials of finite length.)

It is of some interest to classify the factors of ω. Among the obvious factors are finite integers and numbers of the form ω^r where $r < 1$. It is clear from the proof of theorem 8.3 that any number such as $\sum_{i<\alpha} w^{a_i} r_i$ where a_0 is infinitesimal (e.g., $\omega^{\frac{1}{\omega}} + 1$) is a factor. With caution one can take certain products of these such as $(\omega^{\frac{1}{2}}\cdot 7)(\omega^{\frac{1}{\omega}}+1)$. Of course one cannot take all possible products; e.g., $(\omega^{1-\frac{1}{\omega}})(\omega^{\frac{1}{\omega}}+1) > \omega$, so it is certainly not a factor of ω.

The question is whether there exists factors of a more subtle kind. Specifically, can there exist factors such as $\sum_{i<\alpha} \omega^{a_i} r_i$ where $a_0 - a_1$ is not infinitesimal? I feel that this question is of interest in connection with the open problems concerning primes. Although we have no theorems to the effect that an answer to the latter question will imply an answer to any of our questions on primes, I believe that the techniques of the proof will supply insights to study the classification of primes.

Furthermore, the above question leads to algebraic questions of interest which are independent of the theory of surreal numbers. I believe that generalized power series in which exponents may be irrational and lengths may be larger than ω are worthy of study for their own sake. Generalized series do arise in certain subjects such as valuation theory, but as a whole such objects have been neglected in the literature.

We close this chapter by proving a partial result concerning possible factorizations of ω.

<u>Theorem 8.7.</u> ω cannot be expressed in the form cd where $c = \sum_{i<\omega} \omega^{a_i} r_i$ and $d = \sum_{i<\omega} \omega^{b_i} s_i$ are integers, and all the a's and b's are real numbers.

<u>Remark.</u> No restriction on the field of coefficients is needed for the proof. Also we are tacitly assuming that the r's and s's are distinct from 0, i.e. the series are in normal form.

<u>Proof.</u> Suppose $\omega = (\sum_{i<\omega} \omega^{a_i} r_i)(\sum_{i<\omega} \omega^{b_i} s_i)$. By suitable normalization this may be expressed in the form $1 = (1+\sum_{i<\omega} \omega^{a_i} r_i)(1+\sum_{i<\omega} \omega^{b_i} s_i)$ with, of course, not necessarily the same a's, b's, r's and s's.

Since we are considering formal multiplication only we may

replace ω^{-a} by x^a. Also, since c and d are integers the original exponents are positive;, hence in the normalization all the exponents are bounded above by 1. Thus we finally have $1 = (1+\sum x^{a_i} r_i)(1+\sum x^{b_i} s_i)$ where (a_i) and (b_i) are strictly <u>increasing</u> bounded ordinary (i.e. of length ω) sequences of real numbers. We would like to show that this is impossible regardless of the field of coefficients. (It is, of course, trivial to satisfy the above identity if we do not require that (a_i) and (b_i) be bounded.)

Let $\sup(a_i) = \ell$ and $\sup(b_i) = m$. If $\ell < m$ then we can write $x^{m-\ell} = (x^{m-\ell}+\sum x^{a_i+m-\ell} r_i)(1+\sum x^{b_i} s_i)$. Then $\sup(a_i+m-\ell) = \ell+m-\ell = m = \sup b_i$. The equality says that all terms after the first cancel. Thus we finally have the following situation. We have two series $\sum \omega^{c_i} r_i$ and $\sum \omega^{d_i} s_i$ with (c_i) and (d_i) both increasing bounded sequences of real numbers with the same sup ℓ such that all pairs cancel except for the product of the first terms. For convenience we may choose $r_0 = s_0 = 1$. We will now obtain a contradiction.

We now choose an arbitrary but fixed c_i. Let $\delta = \min[c_i-c_{i-1},(c_i-d_j:d_j<c_i)]$. [Since $\sup(c_i) = \sup(d_i)$, $(d_j:d_j<c_i)$ is finite.] $\delta > 0$. Now choose n_0 such that $n \geq n_0 \to d_n > \ell-\delta$.

Now the term $(x^{c_i} r_i)(x^{d_n} s_n)$ for $n \geq n_0$ must cancel in the product. (We may just as well assume that $n_0 \geq 1$, so that the exception consisting of the product of the first terms will not arise.) Hence there must exist a pair (j,m) distinct from (i,n) such that $c_i + d_n = c_j + d_m$. (This is certainly a necessary condition for cancellation.) Therefore $c_i + d_n > c_i + \ell-\delta \geq c_{i-1} + \ell \geq c_k + d_m$ for any $k \leq i-1$ and any m. Hence $j > i$. Furthermore, if d_m were less than c_i we would obtain similarly that $c_i + d_n > c_i + \ell-\delta \geq d_m + \ell \geq c_j + d_m$ for all j. This contradicts the equality. Hence $d_m \geq c_i$.

Now let $\delta' = \min[c_{i+1}-c_i, (d_j-c_i:d_j>c_i)]$. This exists since d_i is increasing. Then $c_i+\ell \geq c_i + d_n = c_j + d_m \geq c_i + \delta'+d_m$. Hence $\ell \geq \delta'+d_m$, i.e. $d_m \leq \ell-\delta'$. Thus if we vary n the set of m's which occur is finite. Similarly, if $d_m > c_i$ we obtain $d_m + \ell \geq c_i + \delta' + \ell \geq c_i + \delta' + d_n = c_j + \delta' + d_m$. Hence $\ell \geq \delta' + c_j$, i.e. $c_j \leq \ell-\delta'$. Thus the set of j's which occur is finite.

We have shown that for $d_m \neq c_i$ these are only finitely many pairs (j,m) such that $c_j + d_m = c_i + d_n$ for any $n \geq n_0$. Hence there exists an n_1 such that for $n \geq n_1$ the only possible pair (j,m) which satisfies $c_j + d_m = c_i + d_n$ must have $d_m = c_i$ and hence $c_j = d_n$. Since i was arbitrary to begin with and also since the same argument can be applied to d_i this shows that $c_i = d_i$ for all i. Furthermore, because of cancellation we must have $r_i s_n = -r_n s_i$ for $n \geq n_1$.

If we begin with $i = 0$, since $r_0 = s_0 = 1$ this tells us that for $n \geq n_1$, $r_n = -s_n$. Now let $i = n_1$. Then for $m \geq n_2$ we have $r_m s_n = -r_n s_m$. Since $r_n = -s_n \neq 0$ this gives $r_m = s_m$.

Hence if $m \geq \max(n_1, n_2)$ then $r_m = s_m$ and $r_m = -s_m$; i.e. $s_m = 0$ which is a contradiction.

Unfortunately, the proof breaks down for lengths other than ω. The limited interest of the theorem is due to the fact that it is the only result we have which gives some restriction to factorizations, thus giving us minimal hope for the existence of non-trivial primes. All our other results give circumstantial evidence against the existence of such primes.

9 GENERALIZED EPSILON NUMBERS

A EPSILON NUMBERS WITH ARBITRARY INDEX

On page 35 in [1] Conway makes some remarks on the possibility of extending the transfinite sequence of epsilon numbers to more general indices, e.g. he gives a meaning to ε_{-1}. He also mentions other interesting surreal numbers. He mentions that the equation $\omega^{-x} = x$ has a unique solution and that there exist various pairs (x,y) satisfying $\omega^{-x} = y$ and $\omega^{-y} = x$. In this chapter we study this systematically. Moreover, we also discuss higher order fixed points, e.g. in the sequence of ordinary ε numbers $\varepsilon_0, \varepsilon_1 \ldots$ there exist α such that $\varepsilon_\alpha = \alpha$ and such α can be parametrized by ordinals. It is interesting that a general elegant theory exists for surreal numbers which have such fixed point properties.

First, we summarize the situation for ordinals. It comes as a surprise to the beginner that although ω^α apparently increases much faster than α, there exist ordinals such $\omega^\alpha = \alpha$. Such ordinals, known as epsilon numbers, can be arranged in an increasing transfinite sequence $\varepsilon_0, \varepsilon_1 \cdots \varepsilon_\alpha$. Not only is this sequence defined for all α but there exist α such that $\varepsilon_\alpha = \alpha$. Furthermore, this construction can be extended indefinitely. Specifically, let us use the notation $\varepsilon_0(\alpha) = \varepsilon_\alpha$ and let $\varepsilon_1(0)$ be the least α such that $\varepsilon_0(\alpha) = \alpha$. Then for every ordinal β, there is an increasing sequence $\varepsilon_\beta(0), \varepsilon_\beta(1), \cdots, \varepsilon_\beta(\omega)$ where $\varepsilon_\beta(\alpha)$ runs through all ordinals which are fixed for every ε_γ with $\gamma < \beta$.

It is possible to do even more by using a kind of "diagonalization." It turns out that the function $\varepsilon_\alpha(0)$ is continuous as a function of α, so there exist α such that $\varepsilon_\alpha(0) = \alpha$. [If $\varepsilon_\beta(\gamma)$ is regarded as a double array, the sequence $\varepsilon_\alpha(0)$ looks more like a first column than a diagonal but it plays the role of a diagonalization. On the other hand, $\varepsilon_\alpha(\alpha)$ is not a continuous function of α and is thus not worth considering.] It is hence possible to form a new double array and then continue as before.

We now turn to surreal numbers. First, we shall deal with epsilon numbers. Although the results are special cases of theorems about general fixed points which we shall prove later, we feel that it is pedagogically reasonable to handle this relatively concrete case first.

Define $\omega_n(a)$ inductively for all positive integers n and surreal numbers a as follows: $\omega_1(a) = \omega^a$ and $\omega_{n+1}(a) = \omega^{\omega_n(a)}$. We are now ready to define ε_b inductively for arbitrary surreal numbers b. Let $b = B'|B''$ be the canonical representation of b. Then $\varepsilon_b = \{\omega_n(1), \omega_n[\varepsilon_{b'}+1]\}|\{\omega_n[\varepsilon_{b''}-1]\}$ where n is an arbitrary positive integer and as usual b' and b" are general elements of B' and B" respectively.

Before stating the basic theorem we look at several examples. The definition gives $\varepsilon_0 = \{\omega_n(1)\}|\phi = \text{l.u.b. } \{\omega_n(1)\}$ which is the ordinary first ε number. We abbreviate ε_0 as ε. Then

$\varepsilon_1 = \{\omega_n(1)\}|\{\omega_n(\varepsilon-1)\}$ which is $\{\omega,\omega^\omega,\omega^{\omega^\omega},\cdots\}|\{\varepsilon-1,\omega^{\varepsilon-1}, \omega^{\omega^{\varepsilon-1}},\cdots\}$. Note that the sequence which comprises the upper set is decreasing although this might run counter to intuition. This is so because $\omega^{\varepsilon-1} \ll \omega^\varepsilon = \varepsilon$ whereas $\varepsilon-1 \sim \varepsilon$. Moreover, once we know that $\omega^{\varepsilon-1} < \varepsilon-1$ the decreasing nature of the sequence follows by induction.

<u>Theorem 9.1</u>. ε_b is defined for all b and satisfies $\omega^{\varepsilon_b} = \varepsilon_b$. ε_b is a strictly increasing function of b and $\varepsilon_b > \omega_n(1)$ for all b and all positive integers n.

<u>Proof</u>. As usual we do this by induction. Since $b' < b''$, $\varepsilon_{b'} < \varepsilon_{b''}$. Hence $\varepsilon_{b'} = \omega^{\varepsilon_{b'}} \ll \omega^{\varepsilon_{b''}} = \varepsilon_{b''}$. Since $\varepsilon_{b''} > \omega_1(1) = \omega$ it is clear that $\varepsilon_{b''}-2 \sim \varepsilon_{b''}$. Therefore $\varepsilon_{b''}-2 > \varepsilon_{b'}$, i.e. $\varepsilon_{b''}-1 > \varepsilon_{b'}+1$. Thus for all positive integers n, $\omega_n(\varepsilon_{b'}+1) < \omega_n(\varepsilon_{b''}-1)$. Also, $\omega_n(1) < \varepsilon_{b''}$. Since both sides have the form ω^a it follows that $\omega_n(1) \ll \varepsilon_{b''} \sim \varepsilon_{b''}-1$. So $\omega_n(1) < \varepsilon_{b''}-1$.

Now $\omega^{\varepsilon_{b'}+1} \gg \omega^{\varepsilon_{b'}} = \varepsilon_{b'} \sim \varepsilon_{b'}+1$. Hence $\omega^{\varepsilon_{b'}+1} > \varepsilon_{b'}+1$.

Similarly $\omega^{\varepsilon_{b''}-1} << \omega^{\varepsilon_{b''}} = \varepsilon_{b''} \sim \varepsilon_{b''}-1$, hence $\omega^{\varepsilon_{b''}-1} < \varepsilon_{b''}-1$. This implies that $\omega_n(\varepsilon_{b'}+1)$ is an increasing function of n and that $\omega_n(\varepsilon_{b''}-1)$ is a decreasing function of n. Hence if m and n are arbitrary positive integers we have $\omega_m(\varepsilon_{b'}+1) < \omega_{m+n}(\varepsilon_{b'}+1) < \omega_{m+n}(\varepsilon_b -1) < \omega_n(\varepsilon_{b''}-1)$. Since $\omega_n(1)$ is certainly an increasing function of n we also have $\omega_m(1) < \omega_n(\varepsilon_{b''}-1)$. We now have exactly what we need to conclude that ε_b is defined.

The fact that $\omega_n(1) < \varepsilon$ is immediate since $\omega_n(1)$ occurs among the lower elements. It is also immediate from the definition that $\varepsilon_b < \omega_1(\varepsilon_{b''}-1) < \varepsilon_{b''}-1 < \varepsilon_{b''}$. Similarly, $\varepsilon_{b'} < \varepsilon_{b''}$. The usual argument using common initial segments shows that ε_b is strictly increasing.

Since the set $\{\omega_n(1)\}$, $\{\omega_n(\varepsilon_{b'}+1)\}$ contains no maximum and the set $\{\omega_n(\varepsilon_{b''}-1)\}$ contains no minimum we know that

$$\omega^{\varepsilon_b} = \{0,\omega^{\omega_n(1)}, \omega^{\omega_n(\varepsilon_{b'}+1)}\}|\{\omega^{\omega_n(\varepsilon_{b''}-1)}\}$$
$$= \{0,\omega_{n+1}(1), \omega_{n+1}(\varepsilon_{b'}+1)\}|\{\omega_{n+1}(\varepsilon_{b''}-1)\} = \varepsilon_b$$

by cofinality. This completes the proof.

<u>Corollary 9.1.</u> The uniformity theorem is valid for ε_b.

<u>Remark.</u> Recall that this means that any representation $b = F|G$ will give us the same result.

<u>Proof.</u> This follows from the usual argument using the inverse cofinality theorem and the cofinality theorem since ε_b is an increasing function.

We generalize the usual definition of epsilon numbers to surreal numbers by defining an epsilon number to be any surreal number such that $\omega^a = a$. We thus have a class of epsilon numbers parametrized by the surreal numbers.

<u>Theorem 9.2.</u> Any epsilon number between $\varepsilon_{b'}$ and $\varepsilon_{b''}$ has ε_b as an initial segment.

<u>Proof.</u> Suppose $\varepsilon_{b'} < \varepsilon < \varepsilon_{b''}$. Since $\omega^\varepsilon = \varepsilon$ it follows that $\varepsilon > 0$. Hence $\varepsilon = \omega^\varepsilon > \omega^0 = 1$. Similarly $\varepsilon > \omega$. We obtain immediately by induction that $\varepsilon > \omega_n(1)$ for all n. Assuming $\varepsilon > \omega_n(1)$ we obtain

that $\varepsilon = \omega^{\varepsilon} > \omega^{\omega_n(1)} = \omega_{n+1}(1)$. Epsilon numbers are in particular powers of ω. Hence $\varepsilon < \varepsilon_{b''} \Rightarrow \varepsilon << \varepsilon_{b''}$. Therefore $\varepsilon < \varepsilon_{b''}-1$. Similarly $\varepsilon_{b'} < \varepsilon \Rightarrow \varepsilon_{b'}+1 < \varepsilon$. Hence $\omega_n[\varepsilon_{b'}+1] < \omega_n(\varepsilon) < \omega_n(\varepsilon_{b''}-1)$. But $\omega_n(\varepsilon) = \varepsilon$ by induction. Hence ε_b is an initial segment of ε since ε satisfies the required inequalities.

Corollary 9.2. If b is an initial segment of c then ε_b is an initial segment of ε_c.

Proof. Let $b = B'|B''$ be the canonical representation. By the lexicographical order $B' < c < B''$. Hence $\varepsilon_{b'} < \varepsilon_c < \varepsilon_{b''}$ so that the result follows from theorem 9.2.

Remark. Note that the proof of theorem 9.2 makes no use of the fact that the representation of b is canonical.

Theorem 9.3. Every epsilon number ε is of the form ε_b for some b.

Proof. Let $\varepsilon = F|G$ be the canonical representation of ε. Also let C be the set of indices of epsilon numbers in F and D the set of indices of epsilon numbers in G. Since $F < G$, it follows that $C < D$. Let $a = C|D$. Now $\varepsilon_C < \varepsilon < \varepsilon_D$ by choice of C and D. Hence, by theorem 9.2 ε_a is an initial segment of ε. On the other hand any epsilon number of the form ε_b which is a proper initial segment of ε must be contained in the set $\varepsilon_C \cup \varepsilon_D$, so it cannot equal ε_a. Hence $\varepsilon_a = \varepsilon$.

Thus the parametrization we have gives us the whole class of epsilon numbers.

Note. Induction is not used in the above proof. We do not need the fact that every epsilon number in $F \cup G$ has the form ε_a for some a.

B HIGHER ORDER FIXED POINTS

We now prove a general fixed point theorem.

Theorem 9.4. Let f be a function from surreal numbers to surreal numbers satisfying the following:

(a) For all a, $f(a)$ is a power of ω.

(b) $a < b \Rightarrow f(a) < f(b)$.

(c) There exist fixed sets C and D such that if $a = G|H$ with G containing no maximum and H no minimum then $f(a) = [C,f(G)]|[D,f(H)]$.

Then there exists a function g which is onto the set of all fixed points of f and which satisfies the above hypotheses with respect to the sets $f_n(C)$ and $f_n(D)$, where n is an arbitrary positive integer and f_n stands for the nth iterate of f.

Remark. Note that ω^a satisfies the hypothesis if we let $C = \{0\}$ and $D = \phi$. Note also that the theorem permits us to obtain higher order fixed points by induction, since the conclusion says that g satisfies the hypothesis. Finally, (c) may be regarded as a generalization of continuity for ordinal functions.

Proof. This is essentially a generalization of the proof of theorem 9.1. We define $g(b)$ inductively as

$g(b) = \{f_n(C),\ f_n[2g(B')]\}|\{f_n(D),\ f_n[\frac{1}{2}g(B")]\}$. (Note that in contrast to the proof of theorem 9.1 we are multiplying and dividing by two rather than adding and subtracting one. This is better in general since multiplying or dividing by two is guaranteed to preserve order of magnitude. Adding one does, not as the example $\omega^{-1} \not\asymp \omega^{-1}+1$ shows. Some functions we consider do have such small values in the range. Anyway, any function which preserves the order of magnitude and satisfies the required inequalities can be used in the proof and it is easy to see by mutual cofinality that the same result is obtained.)

We first show inductively that $g(b)$ is defined for all b, is an increasing function of b, and that $f(g(b)) = g(b)$ for all b.

First, because of condition (a), it follows that $f(a) < f(b) \rightarrow f(a) << f(b)$. Now by the inductive hypothesis $g(b') < g(b")$. Hence $f[g(b')] = g(b') < g(b") = f[g(b")]$. Therefore, by the above remark $f[g(b')] << f[g(b")]$, i.e. $g(b') << g(b")$. Thus $2g(b') < \frac{1}{2}g(b")$. Also, since every element of C is below every element in the range of f by condition (c), it follows that

$c \in C \Rightarrow c < g(b'') \Rightarrow c \ll g(b'') \Rightarrow c < \frac{1}{2}g(b'')$. [Condition (c) is applicable since any surreal a can be represented in the required form by brute force in a trivial manner even if the canonical representation does not satisfy the hypothesis of the condition.] Similarly $d \in D \Rightarrow 2g(b') < d$.

Now if a is a fixed point of f then a is positive since it is a power of ω. Hence $a < 2a$. Therefore $f(a) < f(2a)$, thus $a = f(a) \ll f(2a)$. Therefore $2a \sim a < f(2a)$. Similarly, $f(\frac{1}{2}a) \ll f(a) = a$. Hence $f(\frac{1}{2}a) < \frac{1}{2}a$. Combining this with the fact that $C <$ range $f < D$ we see that if $x \in C \cup 2g(B')$ then $f(x) > x$ and if $x \in D \cup \frac{1}{2}g(B'')$ then $f(x) < x$. It follows by induction that $f_n(x)$ is an increasing function of n if $x \in C \cup 2g(B')$ and a decreasing function of n if $x \in D \; \frac{1}{2}g(B'')$. Since $f_n(x)$ is an increasing function of x, by condition (b) and a trivial application of induction we obtain

$f_n[C \cup 2g(B'')] < f_{n+m}[C \cup 2g(B')] < f_{n+m}[D \cup \frac{1}{2}g(B'')] < f_m[D \cup \frac{1}{2}g(B'')]$. This shows finally that $f(b)$ is defined.

It is immediate that g is increasing by the usual argument using common initial segments. Moreover, in the definition of $g(b)$ the lower terms have no maximum and the upper terms no minimum. Hence condition (c) applies and we obtain

$f[g(b)] = \{C, ff_n(C), ff_n[2g(B')]\} | \{D, ff_n(D), ff_n[\frac{1}{2}g(B'')]\}$ which is $g(b)$ by cofinality.

$g(a)$ is clearly a power of ω for all a since a fixed point of f is <u>a fortiori</u> in the range of f. The uniformity theorem is valid by the usual argument. So if $b = G|H$ then

$g(b) = \{f_n(C), f_n[2g(G)]\} | \{f_n(D), f_n[\frac{1}{2}g(H)]\}$.

Now suppose G has no maximum. We claim that $f_n[2g(G)]$ is mutually cofinal with $g(G)$. One direction is clear since we already know that if x is a fixed point of f then $f(2x) > 2x > x$. On the other hand, consider any element of the form $f_n[2g(x)]$ where $x \in G$. Since G has no maximum there exists $y \in G$ such that $y > x$. Then $g(y) > g(x)$. In fact, since they are both powers of ω, $g(y) \gg g(x)$, hence $g(y) > 2g(x)$. Applying f_n to both sides we obtain $f_n g(y) > f_n[2g(x)]$. Since $g(y)$ is a fixed point of f this says that $g(y) > f_n[2g(x)]$. This is exactly what we need. A similar argument applies to the upper elements. Therefore by the cofinality theorem we obtain that $g(b) = \{f_n(C), g(G)\}|\{f_n(D), g(H)\}$. Thus g satisfies condition (c) with respect to the sets $f_n(C)$ and $f_n(D)$. (Incidentally, note that it is not necessary for n to run through the set of all positive integers. By cofinality any subset containing arbitrarily large integers will work as well.)

To complete the proof of the theorem we still must show that every fixed point of f is in the image of g. We first show that the analogue of theorem 9.2 is valid. In fact, let x be a fixed point between $g(B')$ and $g(B")$. Then $c < x$. Since $g(B') < x$ it follows that $g(B') \ll x$, hence $2g(B') < x$. Therefore $f_n[2g(B')] < f_n(x) = x$ and similarly $f_n(C) < x$. Since the same reasoning applies to the upper elements, it follows from the definition that $g(b)$ is an initial segment of x. Also the analogue of corollary 9.2 as well as the remark following the corollary remain valid. We now verify the analogue of theorem 9.3 which is what we need. As in the earlier argument we let x be a fixed point and let $x = F|G$ be the canonical representation. Let A be the set of all a such that $g(a) \in F$ and B the set of all b such that $g(b) \in G$. Then $A < B$. Let $c = A|B$. Then $g(c)$ is an initial segment of x. However, $g(c) \notin g(A) \cup g(B)$, hence $g(c) \notin F \cup G$. Thus $g(c)$ is not a proper initial segment of x so $g(c) = x$.

Theorem 9.4 allows us to construct higher order fixed points by induction. We begin with any f satisfying the hypothesis, e.g. ω^x. Suppose we call the fixed point function f_1. By induction there exist fixed point functions $\{f_n\}$ for all positive integers n where f_{n+1} is obtained from f_n by the construction in the proof of the theorem. We now indicate how the sequence can be extended to functions with transfinite indices. For this purpose we extend theorem 9.4 to certain ordinal sequences of functions.

Theorem 9.4a. Let f_0 be a function satisfying the hypotheses of theorem 9.4. Then there exist functions f_α for every ordinal α satisfying the hypotheses and such that for $\alpha > 0$, f_α is onto the set of all common fixed points of f_β for $\beta < \alpha$ and satisfies condition (c) with respect to the sets $g(C)$ and $g(D)$ where g runs through all finite compositions of f_β for $\beta < \alpha$.

Outline of Proof. We do this by induction. Specifically we assume that we have functions f_β for all $\beta < \alpha$ satisfying the requirements and show how to construct f_α. Since the proof is similar to the proof of theorem 9.4, to avoid tedious repetition we note only the minor modifications required.

First let us note several consequences of the above given properties of f_β for $\beta < \alpha$. First $f_\beta[f_\gamma(x)] = f_\gamma(x)$ if $\beta < \gamma$. Furthermore, if $\beta < \gamma$ every fixed point of f_γ is a fortiori in the range of f_γ, hence is a fixed point of f_β. This implies for example that if α is a non-limit ordinal $\delta+1$ then the common fixed points of f_β for $\beta < \alpha$ are simply the fixed points of f_δ.

We now look at functions $g = f_{\beta_1} f_{\beta_2} \cdots f_{\beta n}$ more closely. By the above we may assume that $\beta_i \geq \beta_{i+1}$ for all i. Suppose that $f_\beta(x) > x$ for all β. It follows by induction and transitivity that $g(x) > x$ for all g. Let g have the above form with $\beta_1 > \beta_2$ and let h be any composition with all indices less than β_1. Since $f_\beta(x)$ is a fixed point for all f_γ with $\gamma < \beta$, it follows that $hf_{\beta_1}(x) = f_{\beta_1}(x)$. Hence $f_{\beta_1}(x) = hf_{\beta_1}(x) > h(x)$. Now let h_1 also be any composition with all indices less than β_1. Then $f_{\beta_1} h_1(x) > f_{\beta_1}(x) > h(x)$. By induction and transitivity $f_{\beta_1}{}^m\, h_1(x) > f_{\beta_1}{}^n\, h(x)$ if $m > n$. Similar results hold if $f_\beta(x) < x$ for all β. These give us the basic inequalities among all the compositions g for a fixed x.

Now if $x \in C$ or x has the form $2y$ where y is a common fixed point of all f_β then $f_\beta(x) > x$ for all β by our earlier proof. Similarly if $x \in D$ or x has the form $\frac{1}{2}y$ where y is a common fixed point then $f_\beta(x) < x$ for all β. The above inequalities now allow us to imitate the proof of theorem 9.4.

Incidentally, we must be careful in our reasoning with inequalities. For example, assuming $f_\beta(x) > x$ for all β, we easily obtain $f_\beta f_\gamma(x) > f_\beta(x)$. However, we do not obtain $f_\beta f_\gamma(x) > f_\gamma(x)$. In fact, if $\beta < \gamma$ we have equality.

We now define $f_\alpha(b)$ inductively on b as follows. $f_\alpha(b) = \{(g(C), g[2f_\alpha(B')]\} | \{g(D), g[\frac{1}{2}f_\alpha(B'')]\}$. In the case where $\alpha = \delta+1$ the above inequalities show that we obtain cofinal subsets if we consider only g's of the form $f_\delta{}^n$. Thus the formula for $f_\alpha(b)$ is consistent with the one obtained in theorem 9.4 starting with the function f_δ. For general α the proof proceeds as before. The first problem that arises is that we need a g bearing the same relation to an arbitrary pair (g_1,g_2) which f_{m+n} bears to the pair (f_m,f_n) in the proof of theorem 9.4. If $g_1 = f_{\beta_1}{}^m h_1$ and $g_2 = f_{\beta_2}{}^n h_2$, then $g = f^{m+n}_{\max(\beta_1,\beta_2)}$ will work because of the above inequalities. So f_α is defined for all α.

The above inequalities also guarantee that the lower terms in the definition have no maximum and the upper terms no minimum. This allows us to write the following, as we did in the earlier theorem:
$f_\beta[f_\alpha(b)] = \{[g'(c),g'g(c),g'g[2f_\alpha(b')]]\}|\{g'(D),g'g(D),g'g[\frac{1}{2}f_\alpha(B")]\}$
where g' is the subset of g consisting of those products for which all subscripts are less than β. (We are, of course, using the inductive hypothesis on the ordinals.) The above inequalities give us the required cofinality to conclude that this is $f_\alpha(b)$. The rest of the argument is identical to the one used in proving theorem 9.4.

Although the most canonical example of the above is the class of higher order epsilon numbers, we will later investigate another interesting fixed point sequence. At any rate we have obtained a rich supply of exotic surreal numbers.

In the next section we return to a more concrete situation when we study sign sequences of epsilon numbers. We hope that the reader is curious about the sign sequence of ε_{-1}, for example.

C SIGN SEQUENCES FOR FIXED POINTS

As an example let us compute the sign sequence of ε_{-1} directly. Recall that $\varepsilon_{-1} = \{\omega_n(1)\}|\{\omega_n(\varepsilon-1)\}$. $\omega_n(1)$ consists of ω_n pluses. $\varepsilon-1$ consists of ε pluses followed by one minus. $\omega^{\varepsilon-1}$ consists of $\omega^\varepsilon = \varepsilon$ pluses followed by $\omega^{\varepsilon+1}$ minuses. (Note the convenience of thinking in terms of blocks of like signs in computations as referred to in chapter five.) We have here an explicit confirmation of the fact stated earlier that $\omega^{\varepsilon-1} < \varepsilon-1$. $\omega^{\varepsilon+1} = \omega^\varepsilon\omega = \varepsilon\omega$. We can now determine $\omega_n(\varepsilon-1)$ by induction. The number of pluses remains unchanged, i.e. is ε for all n. As for the number of minuses at each stage we premultiply by $\omega^{\varepsilon+1} = \varepsilon\omega$. (Recall that ordinal multiplication is what is relevant here.) Hence $\omega_n(\varepsilon-1)$ has $(\varepsilon\omega)^n$ minuses following the ε pluses. This simplifies to $\varepsilon^n\omega$.

$\{\varepsilon^n\omega\}$ is mutually cofinal with $\{\varepsilon^n\}$. It follows directly from the definition that ε_{-1} consists of ε pluses followed by ε^ω minuses.

We now turn to general epsilon numbers.

For convenience of notation we regard an arbitrary surreal number a as beginning with a_0 pluses followed by b_0 minuses, then a_1 pluses, b_1 minus ... a_α pluses, b_α minuses, etc., i.e. we partition the sequence a into strings of like sign. In this notation a_α, of course, may be 0 if α is a limit ordinal.

<u>Theorem 9.5.</u> a is an epsilon number if and only if $a_0 \neq 0$; all a_α different from 0 are ordinary epsilon numbers satisfying $a_\alpha > \underset{\beta<\alpha}{\text{l.u.b.}}\, a_\beta$; and furthermore b_α is a multiple of $\omega^{a_\alpha\omega}$ for $\alpha_\alpha \neq 0$ and a multiple of $\omega^{c_\alpha\omega}$ where $c_\alpha = \sum_{\beta<\alpha} a_\beta$ for $a_\alpha = 0$.

<u>Proof.</u> We compute ω^a by Corollary 5.1 to the sign sequence formula. The blocks in ω^a correspond to the blocks in a. The αth block of pluses in ω^a has order type $\omega^{c_\alpha+a_\alpha}$ and the αth block of minuses has order type $\omega^{(c_\alpha+a_\alpha+1)}b_\alpha$. $\omega^a = a$ if and only if the following equations are satisfied:

(1) $\omega^{c_\alpha+a_\alpha} = a_\alpha$ for all a such that $a_\alpha \neq 0$ and $\omega^{a_0} = a_0$.

(2) $\omega^{c_\alpha+a_\alpha+1}b_\alpha = b_\alpha$ for all α.

<u>Note:</u> Since the sign sequence formula for a power of ω begins with a plus, special consideration is needed for a_0.

Consider the first equation $\omega^{c_\alpha+a_\alpha} = a_\alpha$. Since $\omega^x \geq x$ for ordinals this can be expressed as a conjunction of two equations:

$$\omega^{a_\alpha} = a_\alpha \text{ and } c_\alpha + a_\alpha = a_\alpha.$$

The first says that a_α is an epsilon number. Since epsilon numbers are additive absorbing, the second condition can be expressed as $c_\alpha < a_\alpha$. In particular, (a_α) is a strictly-increasing sequence. Furthermore as epsilon numbers the a_α are a fortiori powers of ω. Thus we can use the above squeeze argument to deal with $\sum_{\beta<\alpha} a_\beta$. In general, $\sum_{d<\alpha} \omega^d \leq (\text{l.u.b.}_{d<\alpha}\ \omega^d)2$. (We really need the "two" here. For example, consider $1+\omega+\omega^2\cdots+\omega^n\cdots+\omega^\omega$.) Hence $\sum_{\beta<\alpha} a_\beta \leq c_\alpha \leq (\text{l.u.b.}_{\beta<\alpha}\ a_\beta)2$. Since epsilon numbers are multiplicative absorbing the condition $c_\alpha < a_\alpha$ can be replaced by the condition $\text{l.u.b.}_{\beta<\alpha}\ a_\beta < a_\alpha$.

Now let us rephrase condition (2) given that condition (1) is satisfied. Since $c_\alpha + a_\alpha = a_\alpha$ this becomes $\omega^{a_\alpha+1} b_\alpha = b_\alpha$. We now use elementary facts with regard to absorption of ordinals. An ordinal x additively absorbs a if and only if $x \geq a\omega$. Hence an ordinal y multiplicatively absorbs $\omega^{a_\alpha+1}$ if and only if all exponents in the normal form are at least $(a_\alpha+1)\omega = a_\alpha\omega$. (If $a_\alpha = 0$ we apply the above reasoning with c_α replacing a_α.) This completes the proof.

Remark. The above theorem with the aid of theorem 9.2 gives us an immediate evaluation of ε_{-1} since it is clear now that every epsilon number less than ε_0 must begin with ε_0 pluses and $\omega^{\varepsilon_0\omega}$ minuses.

One interesting question that can be asked is the following. For which a is it possible to obtain an infinite sequence (a_i) where $a = \omega^{a_1}$ and for all i $a_i = \omega^{a_{i+1}}$? Epsilon numbers obviously have this property. It may be surprising at first that this property characterizes epsilon numbers since it looks substantially weaker than the equation $a = \omega^a$. However, this follows easily from the sign sequence formula. For ordinals we know that $\omega^x \geq x$ for all x hence the sequence terminates and it is immediate by induction that $\omega^{a_i} = a_i$ implies $\omega^a = a$. In general the length of each block is monotonic decreasing and hence is eventually constant. Since the n at which the length becomes constant a priori may depend on α, a little caution is required. However, we can use induction on α to show that the length has been constant <u>from the beginning</u>. For the sequence of a_α's we can use the same argument we used for ordinals using the fact that the earlier a_α's are constant. The same argument works for the b_α's. To see this note that the above argument for ordinals works for any function f which satisfies $f(x) \geq x$ for all x (the argument makes no use of any special properties of ω^x) so that it applies in particular to functions such as $\omega^{c_\alpha + a_\alpha + 1} x$.

We next obtain a formula for the sign sequence of ε_a. The work is facilitated by the next lemma.

<u>Lemma 9.1</u>. Let f and g be strictly increasing maps from the surreal numbers onto the same class S which preserve the initial segment property. Then $f = g$.

<u>Remark</u>. A function f preserves the initial segment property if for all a and b such that a is an initial segment of b, $f(a)$ is an initial segment of $f(b)$. For example, corollary 9.2 says that the function ε_x has this property.

<u>Proof</u>. As usual we use induction. Let $b = B'|B''$ and suppose $f(x) = g(x)$ for all $x \in B' \cup B''$. Now $g(b) \in S$. Therefore $g(b) = f(c)$ for some c. $g(B') < g(b) < g(B'')$. Therefore $f(B') < f(c) < f(B'')$. Since f is an increasing function $B' < c < B''$. Therefore b is an initial segment of c thus $f(b)$ is an initial segment of $f(c) = g(b)$. Similarly $g(b)$ is an initial segment of $f(b)$. Therefore $f(b) = g(b)$.

Since ε_b is an increasing function which preserves the initial segment property by corollary 9.2 it suffices because of theorem 9.5 to find a sign sequence formula using "good judgment."

<u>Theorem 9.6</u>. Using the notation of theorem 9.5 let $d_\alpha = \sum_{\beta<\alpha} a_\beta$. Then the αth block of pluses in ε_a consists of ε_{d_α} pluses and the αth block of minuses of $(\varepsilon_{d_\alpha})^\omega b_\alpha$ minuses.

<u>Proof</u>. Let $f(a)$ be the function given by the above sign sequence rule. It is immediate from theorem 9.5 that the above is an epsilon number for all a. (Note that in general $\omega^{\varepsilon\omega} = (\omega^\varepsilon)^\omega = \varepsilon^\omega$.) It is also clear

that g is onto the class of all epsilon numbers. In fact, given any ε number the inverse image with respect to f is obtained as follows:

(1) $a_\alpha = d_\alpha - \underset{\beta<\alpha}{\text{l.u.b.}}\, d_\beta$ where ε_{d_α} is the length of the αth block of pluses.

(2) b_α is the unique solution of the equation $(\varepsilon_{d_\alpha})^\omega x = c$ where c is the length of the αth block of minuses.

It is clear that f is increasing and has the initial segment property. This is because roughly speaking f maps larger blocks into larger blocks. Hence $f(a) = \varepsilon_a$.

The above theorem gives us a "concrete" way of looking at the epsilon numbers. The term "concrete" is of course relative since this depends on regarding the ordinary epsilon numbers for ordinals as concrete objects. To illustrate the relativeness of the terms "concrete" and "abstract," we may think of the integer "5" as abstract compared to the phrase "5 apples." On the other hand, an abstract category is a concrete example of an internal category!

We next show that higher order fixed points can be handled in a similar manner, i.e. we desire to express the lengths of the blocks in the sign sequences explicitly in terms of corresponding higher order fixed points for ordinals. It turns out that it is easy to generalize theorems 9.5 and 9.6.

Theorem 9.7. Let f be a function from surreal numbers to surreal numbers. Suppose that g and h are functions on the ordinals such that g is strictly increasing continuous with image in the class of powers of ω and h arbitrary except that it never takes on the value 0. Finally assume that the sign sequence for the function f is given as follows:

(1) The αth string of pluses of f(a) has length $g(d_\alpha)$.

(2) The αth string of minuses of f(a) has length $h(d_\alpha)b_\alpha$ where we are using the notation of theorem 9.6. Then $f(a) = a$ iff for all α, $g(a_\alpha) = a_\alpha$, $a_\alpha > \underset{\beta<\alpha}{\text{l.u.b.}}\, a_\beta$, and b_α is a multiple of $[h(d_\alpha)]^\omega$.

Proof. This is similar to the proof of theorem 9.5. In fact, it is easier in the sense that we are given a "handicap," i.e. the analogue of some of the results obtained on the way in the above proof are contained here in the hypothesis.

$f(a) = a$ if and only if the following equations are satisfied:

(1) $g(d_\alpha) = a_\alpha$.

(2) $[h(d_\alpha)]b_\alpha = b_\alpha$.

As before, equation (1) can be expressed as a conjunction of two equations $g(a_\alpha) = a_\alpha$ and $c_\alpha + a_\alpha = a_\alpha$. The first condition says that a_α is a fixed point. This implies in particular that a_α is a power of ω. As before, with the help of the first condition the second condition can be replaced by $a_\alpha > \operatorname{l.u.b.}_{\beta<\alpha} a_\beta$. (In the earlier proof we did not need the full multiplicative absorbing property. It was enough to absorb 2 multiplicatively.)

The equation $[h(d_\alpha)]b_\alpha = b_\alpha$ is equivalent to the condition $[h(d_\alpha)]^\omega$ divides b_α by the theory of ordinals. This completes the proof.

<u>Remarks</u>. We do not require that the range of h consists only of powers of ω although this does occur in the main examples. Strictly speaking, continuity of g is not used in the proof. Continuity guarantees fixed points for g, hence fixed points for f.

Other variations of the theorem are possible and in fact one will be considered later. At this time we are concerned specifically with higher order epsilon numbers. We now state an analogue of theorem 9.6.

<u>Theorem 9.8</u>. Let f satisfy the hypotheses of theorems 9.4 and 9.7. Then the formula for the sign sequence of the fixed point function f' given by theorem 9.4 is as follows.

The αth block of pluses in $f'(a)$ has length $g'(d_\alpha)$ where g' is the fixed point function of g in the theory of ordinals and the αth block of minuses has length $[hg'(d_\alpha)]^\omega b_\alpha$.

The proof is identical to that of the proof of theorem 9.6. Note now that the conclusion says that f' satisfies the hypothesis of theorem 9.7 to make an induction possible. We let g' play the role of

g and $(hg')^{\omega}$ the role of h. By theorem 9.6 the function ε satisfies the hypothesis of theorem 9.8 so we can use the theorem to determine the sign sequence of ε_n for finite n.

As a convenient reference for computation and recognizing a pattern let us use the diagram $[g,h] \to [g',(hg')^{\omega}]$ which express the parameters for the given function. Then ε_0 corresponds to $[\varepsilon,\varepsilon^{\omega}]$ so ε_1 corresponds to $\{\varepsilon_1,[(\varepsilon_{\varepsilon_1})^{\omega}]^{\omega}\}$. Since the image of ε_1 consists of fixed points of ε by definition, this simplifies to $[\varepsilon_1,(\varepsilon_1)^{\omega^2}]$.

By our earlier remarks we now have a setup for induction which allows us to express the higher order fixed point functions of finite index for surreals in terms of the corresponding functions for ordinals. Thus the pair corresponding to ε_n is $[\varepsilon_n,(\varepsilon_n)^{\omega^{n+1}}]$.

We now outline the continuation to transfinite indices. The pair corresponding to ε_a is $[\varepsilon_a,(\varepsilon_a)^{\omega^{a}}]$. The term ε_a in the pair is clear. The justification of $(\varepsilon_a)^{\omega^{a}}$ follows from the remark that any multiple of ω^{β} for all $\beta < \alpha$ where α is a limit ordinal is necessarily a multiple of ω^{α}, hence a multiple of $x^{(\omega^{\beta})}$ for all $\beta < \alpha$ is necessarily a multiple of $x^{(\omega^{\alpha})}$.

It is nice that such a comparatively explicit description of higher order fixed points exists.

D QUASI ε-TYPE NUMBERS

In this section we introduce other interesting surreal numbers. First, we prove the existence of a solution to the equation $\omega^{-x} = x$. In fact, we let $a_o = 0$ and $a_{n+1} = \omega^{-a_n}$. Then heuristically speaking, (a_n) is an alternating sequence converging to the unique solution of the above equation. We shall also construct other pairs (x,y) such that $\omega^{-x} = y$ and $\omega^{-y} = x$. For convenience let $f(x) = w^{-x}$ and $g(x) = ff(x)$.

<u>Theorem 9.9.</u> $\omega^{-x} = x$ has a unique solution which is obtained as follows: Let $a_o = 0$ and define a_n inductively by $a_{n+1} = \omega^{-a_n}$. Then (a_{2n}) is a strictly increasing sequence, (a_{2n+1}) is a strictly decreasing sequence, and $a_{2n} < a_{2n+1}$ for all n. The unique solution is then

given by $\{a_{2n}\}|\{a_{2n+1}\}$.

Proof. Since ω^{-x} is a decreasing function of x it is immediate that the equation $\omega^{-x} = x$ has at most one solution. Now $f(x) = \omega^{-x}$ is a strictly decreasing function of x, hence $g(x) = ff(x)$ is a strictly increasing function. a_2 is certainly a power of ω hence $g(a_0) = a_2 > 0 = a_0$. So by induction a_{2n} is a strictly increasing sequence. From $a_{2n+2} > a_{2n}$ we obtain $a_{2n+3} = f(a_{2n+2}) < f(a_{2n}) = a_{2n+1}$; therefore a_{2n+1} is a strictly decreasing sequence. Again, a_1 is a power of ω so $a_0 < a_1$. Since g is a strictly increasing function we obtain by induction that $a_{2n} < a_{2n+1}$. It follows that for arbitrary a_{2m} and a_{2n+1} we have $a_{2m} \leq a_{2m+2n} < a_{2m+2n+1} \leq a_{2n+1}$.

Since all even terms are less than all odd terms, $\{a_{2n}\}|\{a_{2n+1}\}$ has meaning. Let $a = \{a_{2n}\}|\{a_{2n+1}\}$. Then $-a = \{-a_{2n+1}\}|\{-a_{2n}\}$. Since the lower terms have no maximum and the upper terms no minimum, we have

$$\begin{aligned}\omega^{-a} &= \{0,\omega^{-a_{2n+1}}\}|\{\omega^{-a_{2n}}\} = \{0,f(a_{2n+1})\}|\{f(a_{2n})\}\\ &= \{a_{2n}\}|\{a_{2n+1}\} = a.\end{aligned}$$

Instead of beginning with 0 one can begin with any a such that $g(a)$ is between a and $f(a)$ and still obtain the same kind of "convergence." Our later work will enable us to easily find examples of a which satisfy the above condition as well as examples which do not.

It is easy to see that g satisfies the hypothesis of theorem 9.4. In fact, conditions (a) and (b) are immediate. Now suppose that $a = G|H$ with G containing no maximum and H no minimum. Then $-a = -H|-G$ hence $\omega^{-a} = \{0,\omega^{-H}\}|\{\omega^{-G}\}$. Therefore $-\omega^{-a} = \{-\omega^{-G}\}|\{0,-\omega^{-H}\}$. Finally, $g(a) = \omega^{-\omega^{-a}} = \{0,\omega^{-\omega^{-G}}\}|\{\omega^0,\omega^{-\omega^{-H}}\} = \{0,g(G)\}|\{1,g(H)\}$. Therefore condition (c) is satisfied with $c = \{0\}$ and $D = \{1\}$.

Recall that for the earlier fixed points D was the empty set, so that it would have been adequate to state condition (c) in the form $f(a) = [C,f(G)]|[f(H)]$ for the immediate applications. The more general form was used in the statement of theorem 9.4 with the present example in mind.

At any rate, we now know that we have a fixed point function g' for g. Note that since $g(x) = ff(x)$, any pair of the form (x,y) where x is a fixed point of g and $y = f(x)$ satisfies $\omega^{-x} = y$ and $\omega^{-y} = x$. If x is a fixed point of g so is $f(x)$. We now show that the indices of x and $f(x)$ are related in a simple manner.

Theorem 9.10. $g'(-b) = f[g'(b)]$.

Proof. By the general formula $g'(0) = [g_n(C)]|[g_n(D)]$. This is precisely $\{a_{2n}\}|\{a_{2n+1}\}$ in the notation of theorem 9.9. Hence $g'(0)$ is the unique solution of $\omega^{-x} = x$. Hence $f[g'(0)] = g'(0) = g'(-0)$ so the theorem is true when $b = 0$. (Although we did not need a separate proof for the case $b = 0$ it is of interest to see it explicitly, especially since the proof is so simple.)

We now need several inequalities before setting up the induction as usual.

We first claim that $2f(x) < f(\frac{1}{2}x) < ff[2f(x)]$ for any fixed point x of g. Since a fixed point x is necessarily larger than 0, we have $\frac{1}{2}x < x$. Hence $f(x) < f(\frac{1}{2}x)$. Since the range of f consists of powers of ω and $2f(x) \sim f(x)$, we even have $2f(x) < f(\frac{1}{2}x)$. Also $2f(x) > f(x)$, hence $f(2f(x)] < ff(x) = x$ since x is a fixed point of $g(x) = ff(x)$. Again since we are dealing with powers of ω we have $f(2f(x)] \ll x$. Therefore $f[2f(x)] < \frac{1}{2}x$ so finally $f(\frac{1}{2}x) < ff[2f(x)]$.

Similarly we obtain the inequality $ff[\frac{1}{2}f(x)] < f(2x) < \frac{1}{2}f(x)$. First, $f(2x) \ll f(x)$ so $f(2x) < \frac{1}{2}f(x)$. Also $\frac{1}{2}f(x) < f(x)$; therefore $f\left[\frac{1}{2}f(x)\right] \gg ff(x) = x$. Hence $f\left[\frac{1}{2}f(x)\right] > 2x$. Finally we obtain $ff\left[\frac{1}{2}f(x)\right] < f(2x)$.

We are now ready for the induction. By theorem 9.4 $g'(b) = \{g_n(0), g_n[2g'(b')]\}|\{g_n(1), g_n\left[\frac{1}{2}g'(b'')\right\}$. As we saw in the proof of the latter theorem, the lower elements have no maximum and the upper elements no minimum, hence

$f[g'(b)] = \{0, fg_n(1), fg_n[\frac{1}{2}g'(b")]\} | \{fg_n(0), fg_n[2g'(b')]\}$. (Note the reversal of sides since $f(x) = \omega^{-x}$.) Since $g(x) = ff(x)$ and $f(0) = 1$ the right-hand side is

$\{g_n(0), fg_n[\frac{1}{2}g'(b")]\} | \{g_n(1), fg_n[2g'(b')]\}$. Also $-b = \{-b"\} | \{-b'\}$ hence

$$g'(-b) = \{g_n(0), g_n[2g'(-b")]\} | \{g_n(1), g_n[\frac{1}{2}g'(-b')]\}$$

$$= \{g_n(0), g_n[2fg'(b")]\} | \{g_n(1), g_n[\frac{1}{2}fg'(b')]\}$$

by the inductive hypothesis.

The inequalities we obtained earlier are just what are needed to check mutual cofinality. First we have $2f(x) < f(\frac{1}{2}x) < ff[2f(x)]$ for any fixed point, in particular for x of the form $g'(b")$. Since g is an increasing function this gives

$g_n[2f(x)] < g_n f(\frac{1}{2}x) < g_{n+1}[2f(x)]$. Similarly from the inequality $ff[\frac{1}{2}f(x)] < f(2x) < \frac{1}{2}f(x)$ we obtain

$g_{n+1}[\frac{1}{2}f(x)] < g_n f(2x) < g_n[\frac{1}{2}f(x)]$. Hence $g'(-b) = f[g'(b)]$.

The fixed points we obtained here are quite different from the earlier kind. Generalized epsilon numbers are all above $\omega_n(1)$ for all n. The present ones are all infinitesimal. They may be regarded as "large infinitesimals." Since they are all squeezed between a_{2n} and a_{2n+1} in the notation of theorem 9.9, it is of interest to determine the first few terms of (a_n) to get an intuitive idea of the size of these fixed points. In fact, $a_0 = 0$, $a_1 = 1$, $a_2 = \frac{1}{\omega}$ and $a_3 = \omega^{-\frac{1}{\omega}}$. Thus a_2 is the canonical infinitesimal, and a_3 being above $\omega^{-\frac{1}{n}}$ for all positive integers n is a large infinitesimal. A study of the sign sequences will enhance the intuition further.

E SIGN SEQUENCE IN QUASI CASE

First, we determine the sign sequence of the unique solution a of $\omega^{-x} = x$ which we feel is a surreal number of special significance (just like the first epsilon number for ordinals). There are essentially two different approaches which can be used for the computation. One can compute the sequence (a_n) inductively and use theorem 9.9 or work directly with the equation $\omega^{-x} = x$. By direct computation the beginning terms of (a_n) are as follows: a_o is the null sequence, $a_1 = (+)$, a_2 consists of a plus followed by ω minuses, a_3 consists of a plus followed by ω minuses and ω^ω pluses, a_4 consists of a_3 followed by $\omega^{\omega+1}\omega^\omega = \omega^{\omega\cdot 2}$ minuses. (Note how the sign sequence so far suggests the heuristic idea of "large infinitesimal.") We now see a pattern emerging which allows us to obtain a_n for all n by induction as well as a. For convenience regard a general surreal number as beginning with c_o pluses d_o minuses, c_1 pluses, d_1 pluses $\cdots$ c_α pluses, d_α minuses, etc.

Theorem 9.11. The unique solution of $\omega^{-x} = x$ has the following sequence: $c_o = 1$; $d_o = \omega$; for each non-negative integer n, $c_{n+1} = \omega^{d_n}$ and $d_{n+1} = (c_{n+1})^2$; and for $\alpha \geq \omega$ both c_α and d_α are zero.

Proof. We first show that the above pattern is followed by all a_n in the sense that a_n is the initial segment consisting of the first n strings of like signs. We already know this for all a_n with $n \leq 4$ and use complete induction on n, i.e. assume that the pattern is followed for all a_m where $m \leq n$. We will show that the pattern remains valid for a_{n+1}.

Now a_n is obtained from a_{n-1} by adding a sequence of like signs. Hence $a_{n+1} = \omega^{-a_n}$ is obtained from $a_n = \omega^{-a_{n-1}}$ by adding a sequence of like signs by the sign sequence formula. Because of the minus in the exponent the signs are opposite to those in the last string of a_n. We now separate the even and odd case to determine the length of the extra string. By the sign sequence formula $c_n = \omega^{\sum_{i=0}^{n-1} d_i}$ and $d_n = (c_n\omega)c_n$. Now, since the square of a power of ω is also a power of ω, it is clear that all c's and d's with the exception of c_o are powers of ω. Also the sequence $c_o, d_o, c_1, d_1, \cdots$ is strictly increasing. Therefore $\sum_{i=0}^{n-1} d_i = d_{n-1}$. Also $c_n \geq \omega^\omega$ if $n \geq 1$. Therefore $c_n\omega c_n$ simplifies to c_n^2. Thus the above becomes $c_n = \omega^{d_{n-1}}$ and

$d_n = c_n^{\,2}$.

There are now two ways to conclude, each of which is immediate. First $\{a_{2n}\}$ and $\{a_{2n+1}\}$ are cofinal sets in the canonical representation of the proposed sign sequence. Thus the sequence is $\{a_{2n}\}|\{a_{2n+1}\} = a$. (It is possibly easier to go back to the definition of $F|G$.) Second, one can ignore the individual a_n and note that the above computation shows directly that the proposed sequence satisfies $\omega^{-x} = x$.

It may seem at first that this method can produce other solutions of $\omega^{-x} = x$. This leads to the scary feeling that something is wrong; however, fortunately the attempt fails and instead the computation leads to information which is consistent with what we have so far. In fact, uniqueness can be checked by sign sequence reasoning by a method which is similar to what was used in the above proof (which, incidentally, is somewhat analogous to the technique of solving differential equations by power series.) c_n and d_n are all uniquely determined. If the sequence for x continues beyond this, i.e. if c_ω or $d_\omega \neq 0$ then ω^{-x} must continue with an opposite sign so ω^{-x} cannot possibly equal x. Thus the uniqueness follows directly from the sign sequence formula.

We next classify the fixed points of $g(x) = \omega^{-\omega^{-x}}$.

Theorem 9.12. x is a fixed point of g if and only if x begins with a and then all c's and d's are epsilon numbers larger than ε_0, each being larger than the least upper bound of the predecessors.

Remark. Note that in contrast to the earlier type of fixed points here the c's and d's have similar restrictions.

Proof. Note first that the length of a is ε_0. This follows easily from a Cantor-Bernstein type argument since it is immediate that $\ell(a) \leq \varepsilon_0$ and that $\ell(a) \geq \varepsilon_0$.

A fixed point of g must begin with a. Conversely, for any sequence x beginning with a the first ω strings of like signs in x and $g(x)$ are alike. Thus it suffices to study the strings of like signs beyond a.

If c_α and d_α are lengths of strings in x let c_α' and d_α' be the corresponding lengths in $f(x) = \omega^{-x}$ (the signs shift of course) and c_α'' and d_α'' be the corresponding lengths in $g(x) = \omega^{-\omega^{-x}}$. (Note that by the sign sequence formula for $\alpha \geq \omega$, the αth string of $g(x)$ correspond to the αth string of x. This is in contrast to the case $\alpha < \omega$ where the αth string in x gives rise to the $(\alpha+1)$st string in $f(x)$.) We now compute c_α'' and d_α''.

First $c_\alpha' = (\omega^{\sum_{\beta<\alpha} d_{\beta+1}})c_\alpha$ and $d_\alpha' = \omega^{\sum_{\beta\leq\alpha} d_\beta}$

Similarly $c_\alpha'' = \omega^{\sum_{\beta\leq\alpha} c_\beta'}$ and $d_\alpha'' = (\omega^{\sum_{\beta<\alpha} c_{\beta+1}'})d_\alpha'$.

It would be messy to express c_α'' and d_α'' in terms of c_α and d_α. Fortunately this is not needed in order to obtain the decisive inequalities.

Suppose the sign sequence has the form described in the statement of the theorem. then we can reason similarly as in the proof of theorem 9.5. By absorption $c_\alpha' = c_\alpha$ and $d_\alpha' = d_\alpha$. (Beware that this does <u>not</u> say that $\omega^{-x} = x$ since all signs are reversed.) Similarly $c_\alpha'' = c_\alpha'$ and $d_\alpha'' = d_\alpha'$. Since the signs have been reversed twice, this shows that $g(x) = x$.

Now suppose $g(x) = x$. Then $c_\alpha'' = c_\alpha$ and $d_\alpha'' = d_\alpha$. (Recall that we are dealing only with $\alpha > \omega$.) From the above formulas $c_\alpha'' \geq \omega^{c_\alpha'} \geq \omega^{c_\alpha}$. Hence $c_\alpha = c_\alpha'' \geq \omega^{c_\alpha}$ thus c_α is an epsilon number. Similarly $d_\alpha = d_\alpha'' \geq d_\alpha' \geq \omega^{d_\alpha}$ hence d_α is also an epsilon number. Furthermore $d_\alpha = d_\alpha'' \geq d_\alpha' \geq \sum_{\beta\leq\alpha} d_\beta$ and $c_\alpha = c_\alpha'' \geq \sum_{\beta\leq\alpha} c_\beta' \geq \sum_{\beta\leq\alpha} c_\beta$. This shows that the c's and d's have the required properties when regarded as separate sequences. However, we also have $c_\alpha = c_\alpha'' \geq c_\alpha' \geq \sum_{\beta<\alpha} d_{\beta+1}$ and $d_\alpha = d_\alpha'' \geq \sum_{\beta<\alpha} c_{\beta+1}' \geq \sum_{\beta<\alpha} c_{\beta+1}$. Finally, if we take into account the terms in a we obtain that all c_α's and d_α's are larger than ε_0. This is enough to complete the proof of the theorem.

We can now continue in a manner which is similar to what we did for ordinary epsilon numbers except that now things are easier because of the similar treatment of pluses and minuses. We now no longer discriminate against the poor minuses but use cumulative sums. Thus, regardless of whether the αth string of x consists of pluses or minuses, the αth string of $g'(x)$ beyond a is made up of the same sign and has length ε_{b_α}, where b_α is the total length of the sequence of x up to and including the αth string.

The fixed points we obtained seem interesting because of their contrast with epsilon numbers. Higher order fixed points may be constructed by theorem 9.4, but this does not seem to be interesting enough to pursue in much detail. It suffices to note that we need a variation of theorem 9.7. On the one hand we use cumulative sums rather than d_α and treat pluses and minuses alike so a function h is not needed. On the other hand, we need a preliminary sequence such as our element a. A fixed point of f necessarily begins with the juxtaposition $[a, f_1(a), f_2(a), \cdots]$. Finally, to guarantee a fixed point one must make sure that the length of the latter sequence gets absorbed when added to the fixed points of g to ensure that we obtain a fixed point of f.

10 EXPONENTIATION

A GENERAL THEORY

As was mentioned in the introductory chapter, Kruskal discovered that a theory of exponentiation for the surreal numbers is possible. Taking advantage of his hints I discovered that an elegant natural theory does exist, i.e. exp x can be defined in a uniform way for all surreal numbers x and it has the properties that are expected of an exponential function. Note that the function ω^x is not suitable as an exponential function even though the theory in chapter five makes this notation convenient. For example, it is certainly not onto since no two numbers in the range have the same order of magnitude. (The word "exponent" used in the past is a convenient abuse of language.)

Although we begin with a unified definition of exp x the subject breaks up naturally into three cases.

(a) x is real,

(b) x is infinitesimal,

(c) x is purely infinite (i.e. all "exponents" in the normal form of x are positive).

The unified form is somewhat complicated to deal with, whereas the theory simplifies in each of the above cases for different reasons. (Note that any surreal number is uniquely a sum of three numbers each of which satisfies one of these cases.) Case (c) is the only one which is worthy of a substantial discussion. In case (a) it suffices to show that the unified definition is consistent with the usual one and in case (b) that the unified definition is consistent with the result by formal expansion in the spirit of chapter five.

Although the notation "ω^x" will continue to be used with its original meaning there should be no danger of confusion since the exponential function will consistently be written as exp x. This is because it is obvious how to define general exponentiation in terms of exp x and its inverse function thus there is no need for us to consider expressions such as a^b in particular "ω^b" in the sense of exponentiation.

The motivation for the definition of exponentiation (as it was for multiplication) is based on suitable inequalities satisfied by ordinary real numbers. Usually the construction of a definition requires some ingenuity; one must include enough inequalities to obtain the desired result and avoid circular reasoning. We recall, for example, that for ordinary real numbers we have the implication

$$(a_1 < a) \wedge (b_1 < b) \longrightarrow ab_1 + a_1b - a_1b_1 < ab.$$

Such implications make the definition of multiplication possible. Recall also that in the study of reciprocals in chapter three we could not get by with the obvious inequalities but we were forced to use some subtlety instead.

In searching for a definition of exponentiation similar difficulties occur. It is natural to attempt to use Taylor series approximations. For example, if x is positive, $1+x+\frac{x^2}{2!}\cdots\frac{x^n}{n!}$ is a lower approximation to exp x. However, $\{0,1+1+\frac{1}{2!}\ldots\frac{1}{n!}\}|\phi = 3 \neq \exp 1$ so this is not adequate. Also the sets $\{1+\omega+\frac{\omega^2}{2!}\cdots\frac{\omega^n}{n!}\}$ and $\{1+\frac{(\omega\cdot 2)^2}{2!}\ldots\frac{(\omega\cdot 2)^n}{n!}\}$ are mutually cofinal so that caution is required in framing a definition to guarantee that $\exp\omega \neq \exp(\omega\cdot 2)$. Actually the idea of using Taylor series does work as long as one uses proper refinements.

We use the notation $[x]_n$ for $1+x+\frac{x^2}{2!}\cdots\frac{x^n}{n!}$ for an arbitrary surreal number x. For negative x we will be interested only in odd n for which $[x]_n$ is positive. Thus when we use the notation $[x]_{2n+1}$ when x is negative, it is understood that $[x]_{2n+1}$ is positive. This awkwardness when x is negative is a price that is paid for unification. Note that in case (b) all $[x]_{2n+1}$ are positive and in case (c) we have the other extreme where no $[x]_{2n+1}$ is positive. Thus we have the ironic fact that case (a) which we already know all about from elementary analysis is the one that causes the greatest awkwardness when combined with the other cases! This is why we prefer as soon as possible to split the subject into the three parts according to the different cases. On the other hand, although the theory can be developed by considering each case separately from the beginning, we prefer not to do so since the subject then takes on an ad hoc form with definitions that should really be theorems. In fact, the whole beauty of the subject of surreal numbers from the very beginning lies in the possibility of making uniform definitions which reduce to what is desired in various special cases.

As in the past we use the canonical representation of an element x to define $\exp x$ inductively as follows:

$$\exp x = \{0, (\exp x')[x-x']_n, (\exp x'')[x-x'']_{2n+1}\} | \{\frac{\exp x''}{[x''-x]_n}, \frac{\exp x'}{[x'-x]_{2n+1}}\}.$$

The restriction of $[x]_{2n+1}$ to positive values for the upper elements is essential in order to have any hope for the above definition to work since 0 is a lower element. On the other hand, by cofinality, no harm is done if negative values are included among the lower elements. (Of course, no good is done either!) Also, it turns out that expressions involving $[x]_n$ for negative x and even n are redundant. On the other hand, examples such as ones we mentioned earlier suggest that we need all the terms we have.

Before proving that $\exp x$ is defined and investigating its properties, we mention several needed inequalities among elements of the form $[x]_n$.

(1) For positive, x,y we have $[x]_n[y]_n \geq [x+y]_n$ and $[x+y]_{2n} \geq [x]_n[y]_n$.

The above inequalities depend only on ordered field properties and follow immediately from the binomial theorem. Also we note the obvious fact that $[x]_n$ is an increasing function of n.

For negative x the situation is more complicated. Anyway, for infinite x there are no $[x]_{2n+1}$ to consider, so the problem reduces to the case where $x = r+\varepsilon$ where r is real and ε is infinitesimal.

First, for real r we can apply plain ordinary analysis. Since $1-r+\frac{r^2}{2!}\ldots \to e^{-r} > 0$ there certainly exists n such that $1-r+\frac{r^2}{2!}\ldots-\frac{r^{2n+1}}{(2n+1)!} > 0$. Necessarily $r < 2n+1$. Hence it follows that if n_0 is the least n for which the latter expression is positive, $[r]_{2n+1}$ is defined precisely when $n \geq n_0$, $[r]_{2n+1}$ is an increasing function of n and $\lim[r]_{2n+1} = e^{-r}$. Since $e^{x+y} = e^x e^y$, it is a trivial exercise in analysis to show that

$$\forall m\ \exists n_0 (n \geq n_0 \rightarrow [x]_{2n+1}[y]_{2n+1} \geq [x+y]_{2m+1})$$

and

$$\forall m\ \exists n_0 (n \geq n_0 \rightarrow [x+y]_{2n+1} \geq [x]_{2m+1}[y]_{2m+1}).$$

Since $[-x]_{2n+1} < e^{-x}$ and $[x]_{2n+1} < e^x$, clearly $[-x]_{2n+1}[x]_{2n+1} < 1$ for $x > 0$.

For infinitesimal ε the work simplifies since we can reason purely algebraically using orders of magnitude. Clearly $[\varepsilon]_{2n+1}$ is defined for all n and is increasing.

For negative infinitesimals x and y we also have the inequalities $[x]_{2n+1}[y]_{2n+1} \geq [x+y]_{2n+1}$ and $[x+y]_{4n+3} \geq [x]_{2n+1}[y]_{2n+1}$. These follow from the binomial theorem by reasoning with orders of magnitude. In both inequalities $2n+2$ is the lowest total degree of terms occurring in the left and not on the right, and all such terms have positive coefficients. Also, $[-\varepsilon]_{2n+1}[\varepsilon]_{2n+1} < 1$ for $\varepsilon > 0$. This follows from the fact that on the left-hand side all terms involving ε^i beyond $\varepsilon^0 = 1$ for $i \leq 2n+1$ cancel by the binomial theorem but the coefficient of ε^{2n+2} is negative. Furthermore $1 - [-\varepsilon]_{2n+1}[\varepsilon]_{2n+1} \sim \varepsilon^{2n+2}$; hence $\frac{1}{[-\varepsilon]_{2n+1}} - (\varepsilon)_{2n+1} \sim \varepsilon^{2n+2}$ since $[-\varepsilon]_{2n+1} \sim 1$.

Finally, we consider a finite element of the form $r+\varepsilon$ with r real and ε infinitesimal such that $r > 0$. Then $r+\varepsilon > 0$. Hence $[-(r+\varepsilon)]_{2n+1}$ differs from $[-r]_{2n+1}$ by an infinitesimal. If $[-r]_{2n+1}$ is positive so is $[-(r+\varepsilon)]_{2n+1}$. Similarly, if $[-r]_{2n+1}$ is increasing for $n \geq n_0$ so is $[-(r+\varepsilon)]_{2n+1}$. A petty nuisance is caused by the possibility that $[-r]_{2n+1}$ equals zero for some n. However, we can use the above argument for real r for the improper inequality $1-r + \frac{r^2}{2!} \cdots \frac{r^{2n+1}}{(2n+1)!} \geq 0$ and still obtain an n_0. $[-r]_{2n+1}$ and therefore $[-(r+\varepsilon)]_{2n+1}$ will still be increasing for $n \geq n_0$. Thus there may be an n_0 for which $[-(r+\varepsilon)]_{2n+1}$ is considered but not $[-r]_{2n+1}$ but by the above this can happen for at most one n_0 and no harm is done.

The results relating $[x]_{2n+1}$, $[y]_{2n+1}$, and $[x+y]_{2m+1}$ still remain valid. Since $[x]_{2n+1}$ is a strictly increasing function of n for all x, the improper inequalities in the above theorems for real x and y can be replaced by proper inequalities (by making n_0 larger if necessary). The results for numbers of the form $r+\varepsilon$ then follow immediately from the results for the real parts r.

Again we have $[-(r+\varepsilon)]_{2n+1}[r+\varepsilon]_{2n+1} < 1$.

This gives us the basic properties of $[x]_{2n+1}$ for negative x. For the sake of symmetry between the cases for positive and negative x and for the sake of uniformity, all of our results relating $[x]_n$, $[y]_n$ and $[x+y]_n$ can be expressed in the form given for negative real numbers, i.e. we have no need for the explicit value of n_0 given in several of the cases. What really matters for our purpose is what happens for sufficiently large n.

Now that we are armed with the needed inequalities, we are ready to prove the basic theorem.

<u>Theorem 10.1</u>. $\exp x$ is defined for all x. Furthermore, if $x < y$ then $\exp x\,[y-x]_n < \exp y$ and $\exp y\,[x-y]_{2n+1} < \exp x$ for all positive integers n. Also $\exp x > 0$.

<u>Proof</u>. Recall that $\exp x = \{0, \exp x'[x-x']_n, (\exp x'')[x-x'']_{2n+1}\} \mid \{\frac{\exp x''}{[x''-x]_n}, \frac{\exp x'}{[x'-x]_{2n+1}}\}$. We use induction as usual. First we must show that the lower terms are below the upper terms, so that $\exp x$ is defined.

The inequalities involving 0 are immediate in view of our restriction that all the terms of the form $[x'-x]_{2n+1}$ are positive, and the inductive hypothesis that $\exp x'$ and $\exp x''$ are positive.

We now compare terms of the form $\exp x'[x-x']_m$, and $\frac{\exp x''}{[x''-x]_n}$. First suppose $m = n$. Choose p so that $[x''-x']_p \geq [x''-x]_m[x-x']_m$. By the inductive hypothesis $\exp x'[x''-x']_p < \exp x''$. Therefore $\exp x'[x''-x]_m[x-x']_m < \exp x''$, i.e. $\exp x'[x-x']_m < \frac{\exp x''}{[x''-x]_m}$. Since $[x]_n$ is an increasing function of n, the result for general m and n

follows immediately. In fact,

$$\exp x'[x-x']_m \le \exp x'[x-x']_{\max(m,n)} < \frac{\exp x''}{[x''-x]_{\max(m,n)}} \le \frac{\exp x''}{[x''-x]_n}.$$

Exactly the same argument shows that $\exp x''[x-x'']_{2m+1} < \frac{\exp x'}{[x'-x]_{2n+1}}$. (In fact, we deliberately ignored the fact that p can be taken to equal $2m$ in the above proof in order to make it possible to use the same argument in both cases.

Note that we also have $[x'-x]_{2m+1}[x-x']_n < 1$ for arbitrary positive integers m and n by an argument similar to one used above. Hence $\exp x'[x-x']_n < \frac{\exp x'}{[x'-x]_{2m+1}}$. However, this is not enough. We need an inequality of the form $\exp x_1' [x-x_1'] < \frac{\exp x_2'}{[x_2'-x]_{2m+1}}$ where x_1' and x_2' are two arbitrary lower elements not necessarily equal. It turns out that we can use a device which is similar to what we used for different subscripts m and n which will also be important later; in fact, it is the same cofinality idea which we used, for example, in our development of multiplication in chapter three.

Specifically, we desire to prove the following: Let $x_1' < x_2'$. Then $\forall m\, \exists n\{\exp x_1'[x-x_1']_m \le \exp x_2'[x-x_2']_n\}$. In fact, choose n so that $[x-x_2']_n[x_2'-x_1']_n \ge [x-x_1']_m$. Then $\exp x_2'[x-x_2']_n > \exp x_1'[x_2'-x_1']_n[x-x_2']_n > \exp x_1'[x-x_1']_m$, where we used the inductive hypothesis in the first inequality. Similarly, we have $\forall m\, \exists n \left[\frac{\exp x_2'}{[x_2'-x]} \le \frac{\exp x_1'}{[x_1'-x]_{2m+1}}\right]$. Hence, if $x_3' = \max(x_1',x_2')$, then for a sufficiently high p we have

$$\exp x_1'[x-x_1']m \le \exp x_3'[x-x_3']_p \le \frac{\exp x_3'}{[x_3'-x]_{2p+1}} \le \frac{\exp x_2'}{[x_2'-x]_{2n+1}}.$$

Similarly, if $x_2'' < x_1''$ we have the following inequalities:

$$\forall m\, \exists n[\exp x_1''[x-x_1'']_{2m+1} \le \exp x_2''[x-x_2'']_{2n+1}]$$

and

$$\forall m\, \exists n \left[\frac{\exp x_2''}{[x_2''-x]_n} \le \frac{\exp x_1''}{[x_1''-x]_m}\right]$$

so that we obtain $\exp x_1''[x-x_1'']_{2m+1} \leq \frac{\exp x_2''}{[x_2''-x]_n}$.

Thus we finally have what we need to conclude that $\exp x$ is defined. In addition, the latter inequalities give us important cofinality results; e.g. if $x_1' > x_2'$ we can dispense with x_2' in computing $\exp x$.

Since 0 is a lower element in the definition of x certainly $\exp x > 0$. The inequality $\exp x[y-x]_n < \exp y$ for $x < y$ is immediate from the definition if either y or x is an initial segment of the other. Otherwise we can use the same technique as in chapter three by considering the common initial segment of x and y. Let $x < z < y$ where z is that common initial segment. Now $\forall n \exists p[z-x]_p[y-z]_p > [y-x]_n$. Hence

$\exp y > \exp z[y-z]_p > \exp x[z-x]_p[y-z]_p > \exp x[y-x]_n$. Similarly $\exp y\,[x-y]_{2n+1} < \exp x$. This completes the proof.

<u>Corollary 10.1</u>. The uniformity theorem is valid for exponentiation.

<u>Proof</u>. This is immediate from the inequalities obtained in the proof of the above theorem using as usual the inverse cofinality and cofinality theorems.

We are now ready to study the main cases separately. For real r let us temporarily use the notation e^r for the ordinary exponential function.

<u>Theorem 10.2</u>. $\exp r = e^r$ for real r.

<u>Proof</u>. Note first that $\exp 0 = \{0\}|\phi = 1$ which is a good start although we have a long way to go!

Since all proper initial segments of reals are reals (in fact dyadic) we may use induction. Hence we have

$\exp r = \{0, e^{r'}(r-r')_n, e^{r''}[r-r'']_{2n+1} \,|\, \frac{e^{r''}}{[r''-r]_n}, \frac{e^{r'}}{[r'-r]_{2n+1}}\}$. We use

cofinality to show that this expression gives e^r. Recall that e^r, like any real number, can be expressed in the form $\{e^r-\varepsilon\}|\{e^r+\varepsilon\}$ where ε is positive real. It is immediate from our earlier inequalities that e^r satisfies the betweenness property. For example,

$e^{r'}[r-r']_n < e^{r'}e^{r-r'} = e^r$. Cofinality is also easy. If $r > 0$ then 0, which is one possible value of r', gives rise to lower elements $[r]_n$ and upper elements $\frac{1}{[-r]_{2n+1}}$. Since $\lim_{n\to\infty} [r]_n = \lim_{n\to\infty} \frac{1}{[-r]_{2n+1}} = e^r$, these elements are cofinal in the above expression for e^r. Similarly, if $r < 0$, we can use 0 as a possible value of r''. This completes the proof.

Remark. Note that we used the fact that lower terms in r contribute to upper terms in $\exp r$ and vice versa. Also, in this case the only initial segment we needed was 0. This is not true in general although cofinality results permit us to cut down to some extent on the set of x' and x'' needed to compute $\exp x$. [For example, as we saw earlier the use of $x' = 0$ only would not distinguish $\exp \omega$ and $\exp(\omega\cdot 2)$.]

For infinitesimal x there is a different approach. In this case the formal sequence $1+x+\frac{x^2}{2!}+\ldots$ has meaning by the results of chapter five. This is purely algebraic in contrast to the analytic spirit for real x. Again we use the temporary notation e^x, this time using it for the formal sum.

Theorem 10.3. $\exp x = e^x$ for infinitesimal x. More generally, if x has the form $r+\varepsilon$ where r is real and ε is infinitesimal, then $\exp(r+\varepsilon) = e^r e^\varepsilon$.

Proof. Recall that every finite number x can be expressed uniquely in the form $r+\varepsilon$ where r is real and ε is infinitesimal. Since we already have theorem 10.2 we may assume that $\varepsilon \neq 0$. Note also that e^ε has real part one and in general $[r+\varepsilon]_n$ is finite and has real part $[r]_n$.

Since all initial segments of finite surreals are finite, we may use induction. We show first that by cofinality that in the representation of $\exp x$ we can restrict ourselves to initial segments of the form $r+\delta$ where δ is infinitesimal. (Note that $r+\varepsilon$ might have no upper segments of the above form; e.g., $1+\frac{1}{\omega}$ has only numbers such as $1+\frac{1}{2^n}$ as upper segments, so that this does not follow immediately from our earlier cofinality results.) The basic idea is that r is an initial segment of $r+\varepsilon$ and regardless of whether ε is positive or negative, terms containing exp r contribute to both upper and lower terms of $\exp(r+\varepsilon)$.

Suppose $y = s+\delta$ is a lower initial segment of $x = r+\varepsilon$ where $s < r$ (necessarily $\delta = 0$ as is easily seen, but this fact is irrelevant for our purpose). By the inductive hypothesis and our earlier remarks the real part of $\exp x[x-y]_n$ is $\exp s[r-s]_n$ which is less than e^r. Similarly the contributions of y to the upper elements all have real part larger than e^r. On the other hand, the contributions of r such as $e^r[\varepsilon]_n$ all have real part e^r. Since the same reasoning applies to upper terms this proves the desired cofinality.

Since all the upper and lower terms remaining in the representation of exp x now have essentially the form $e^r e^{\delta_1}[\delta_2]_n$ it remains to study expressions such as $e^{\delta_1}[\delta_2]_n$.

First, the identity $e^{x_1}e^{x_2} = e^{x_1+x_2}$ for infinitesimals x_1 and x_2 follows from the identity for formal power series. Also, by reasoning with formal power series we have $[x]_n < e^x$ and $[-x]_{2n+1} < e^{-x}$.

Now if $r+\delta$ is a lower initial segment of $r+\varepsilon$ then we have terms such as $\exp(r+\delta)[\varepsilon-\delta]_n$ in the representation of $\exp(r+\varepsilon)$. By the inductive hypothesis this is $(\exp r)e^\delta[\varepsilon-\delta]_n$, which is less than $(\exp r)e^\delta e^{\varepsilon-\delta} = (\exp r)e^\varepsilon$.

A similar argument for the other terms in the representation of $\exp(r+\varepsilon)$ shows that $(\exp r)e^\varepsilon$ satisfies the betweenness condition for these terms. Thus to complete the proof it suffices to show that the terms are cofinal in a representation of $(\exp r)e^\varepsilon$. We use the standard representation dealt with in chapter five, i.e. if $(\exp r)e^\varepsilon$ has the form $\sum_{i<\alpha} \omega^{a_i} r_i$ then we have

$(\exp r)e^\varepsilon = \{\sum_{i<\beta} \omega^{a_i} r_i - \omega^{a_\beta}\delta\} | \{\sum_{i<\beta} \omega^{a_i} r_i + \omega^{a_\beta}\delta\}$ where $\beta < \alpha$ and δ is positive real.

Suppose x has the form, $\sum_{i<\alpha} \omega^{a_i} r_i$. Then every exponent b in $(\exp r)e^\varepsilon$ is a finite sum of a's. Let $b = \sum_{j=1}^{n} a_{i_j}$. If $a_\beta = \min(a_{i_j})$ then $b \geq na_\beta$. Now $y = \sum_{i<\beta} \omega^{a_i} r_i$ is an initial segment of $\sum_{i<\alpha} \omega^{a_i} r_i$ and hence participates in obtaining upper and lower terms to $\exp x$. Let $z = x-y = \sum_{i\geq\beta} \omega^{a_i} r_i$.

Suppose first that $z > 0$. Then $\exp y[z]_n$ is a lower term in the representation of $\exp x$. Now $(\exp y)e^z - \exp y[z]_n = (\exp y)[\frac{z^{n+1}}{(n+1)!} + \cdots]$. The first term in $\exp y$ has exponent 0. Hence the first exponent in the latter expression is $(n+1)a_\beta$ which is smaller than na_β. (As a technicality, in case all a_{i_j} are zero, we can choose β to be 1 so that $a_\beta < 0$.) Hence $(\exp y)e^z$ and $\exp y[z]_n$ agree in all terms up to and including the terms with exponent na_β, _a fortiori_ the terms with exponent b.

For $z < 0$ the argument is identical except for the fact that we consider only odd n.

Also note that in general if two expressions $\sum_{i<\beta} \omega^{b_i} s_i$ agree in all terms up to and including exponent c, then so do their reciprocals since it is clear that by formally taking inverses terms with given exponents cannot give rise to terms with earlier exponents. Therefore the above argument applies equally well to expressions such as $\frac{(\exp y)}{e^z}$ and $\frac{\exp y}{[z]_n}$.

If $y = r+\delta$ then by the inductive hypothesis $\exp y = (\exp r)e^\delta$. Hence $(\exp y)e^z = (\exp r)e^{y-r}e^z = (\exp r)e^{y-r+z} = (\exp r)e^\varepsilon$. The same argument applies to upper elements. Thus we have shown that for any given exponent a_β occurring in $(\exp r)e^\varepsilon$ there are terms in the representation of $\exp(r+\varepsilon)$ which agree with $(\exp r)e^\varepsilon$ for all terms up to and including the terms with exponent a_β. By the lexicographical order such terms are cofinal in the above standard representation of of $(\exp r)e^\varepsilon$. This completes the proof.

We now know that $\exp x$ agrees with what is expected for finite x. Also $\exp x$ is clearly strictly increasing and satisfies $\exp(x+y) = \exp x \exp y$. This is clear from theorem 10.3 and the corresponding identities for reals and infinitesimals separately.

For real x we have the function $\ln(1+x)$ which for sufficiently small x may be expressed as $1-x+\frac{x^2}{2} \cdots$. Then

$\exp \ell n(1+x) = 1+x$. This identity is valid for formal series, hence it is applicable to infinitesimal surreal numbers, i.e. it is possible to define a log function in the natural way. For any infinitesimal $\exp \ell n(1+\varepsilon) = 1+\varepsilon$. This shows in particular that the function x maps the infinitesimals onto the class of all surreal numbers of the form $1+\varepsilon$ where ε is infinitesimal. Combining this with the behaviour of exp for ordinary real numbers, we see that exp maps the finite numbers into the class of all positive finite non-infinitesimal numbers.

This essentially completes the theory for the finite case. There were no surprises. On the contrary, the fact that the unified definition agrees with what is expected gives some philosophical justification for using it for the infinite case where we lack an a priori alternative. This is the realm where exotic results occur and, as mentioned earlier, the case of main interest.

From now on we would like to limit our study to the case where x is purely infinite, i.e. where all exponents in the normal form of x are positive. However, before doing so we need a "piecing together" result.

Theorem 10.4. If $x = y+z$ where y is purely infinite and z is finite, then $\exp x = \exp y \exp z$.

Proof. We again use induction. Note that if x has the form $\sum_{i<\alpha} \omega^{a_i} r_i$, then y necessarily has the form $\sum_{i<\beta} \omega^{a_i} r_i$ and z the form $\sum_{i\geq\beta} \omega^{a_i} r_i$. Hence y is an initial segment of x. We use an argument similar to the one used in the proof of theorem 10.3 to show that in the representation of $\exp x$ we can restrict ourselves to initial segments of the form $y+u$ where u is finite.

Regardless of whether z is positive or negative, terms containing $\exp y$ contribute to both upper and lower terms in $\exp(y+z)$. Now suppose $x-x_1$ is positive infinite. Consider a term $\exp x_1[x-x_1]_n$. Note that $[x-x_1]_{n+1} \gg [x-x_1]_n$ by elementary reasoning with orders of magnitude. Therefore $\exp x > \exp x_1[x-x_1]_{n+1} \gg \exp x_1[x-x_1]_n$. A similar inequality applies if $x-x_1$ is negative infinite. On the other hand, y contributes terms such as $\exp y[z]_n$ or $\exp y[z]_{2n+1}$ if z is negative. Now $[z]_n \geq 1$ if z is positive. If z is negative we can certainly choose n so that $[z]_{2n+1}$ is not infinitesimal. In fact, by our earlier discussion $[z]_{2n+1}$ can be infinitesimal for at most one value of n for each z. $[z]_{2n+1}$ is certainly finite. Hence such a term is of the same order of magnitude as $\exp y$. It follows from the above and the inequalities in the statement of theorem 10.1 that $\exp y \sim \exp x$. Thus y contributes a lower term in the representation of $\exp x$ of the same order of magnitude as $\exp x$. This is thus larger than $\exp x_1[x-x_1]_n$ by our earlier inequality.

Since a similar argument applies to upper terms this proves the cofinality.

Now the sign sequence of an element of the form $y+u$ is the sign sequence of y followed by the sign sequence of u. This follows from the fact that all exponents in the normal form of y begin with a plus whereas those in u do not, so that by the sign sequence formula y does not contribute to minuses which are ignored in the sign sequence for $y+u$. Hence the initial segments of $y+z$ are of two types: proper initial segments of y, and y followed by an initial segment u of z.

Since y is an initial segment of $y+w$ for any finite w, $y+w$ is certainly not a proper initial segment of y, so the first type cannot have that form. On the other hand, the second type consists precisely of all numbers of the form $y+v$ where v is an initial segment of z. This says that our cofinality result restricts the initial segments considered precisely to those of the second type, i.e. those of the form $y+u$ where u is an initial segment of z.

Thus the formula for $\exp x = \exp(y+z)$ involves terms such as $\exp(y+u)[z-u]_n$ which by the inductive hypothesis is $\exp y \exp u[z-u]_n$. $\exp u[z-u]_n$ is, of course, a typical term used in the computation of $\exp z$. Similar remarks apply to the other terms so that we finally obtain that $\exp(u+z)$ may be represented as $\{\exp y\, F\}|\{\exp y\, G\}$ where $F|G$ is the representation used in the definition of $\exp z$. ($\exp y\, F$ stands for the set of all products of $\exp y$ with elements of F and similarly for $\exp y\, G$.) We must prove that $\exp y \exp z = \{\exp y\, F\}|\{\exp y\, G\}$ which heuristically is a kind of distributivity. In general there certainly is no such "distributive law" for multiplication. In the present case we make use of the special known properties of y, z, F, and G.

First, since 0 is a lower element of exp y then exp y F and exp y G appear among the terms in the formula for the product exp y exp z. It therefore suffices to show that these particular terms are cofinal with respect to all other terms in the formula for this product.

For this purpose we need some information regarding orders of magnitude. For convenience, let $H|K$ be the representation of exp y used in the definition. Since y is purely infinite, any proper initial segment z of y satisfies $|y-z|$ is infinite. If $z < y$ we have a typical term of the form $\exp y[y-z]_n$. As we already noted, $\exp z[y-z]_n \ll \exp z[y-z]_{n+1} < \exp y$. Similarly if $z > y$ we have $\frac{\exp z}{[z-y]_n} \gg \frac{\exp z}{[z-y]_{n+1}} > \exp y$. Also recall that for negative infinite x there are no terms of the form $[x]_{2n+1}$. Thus we have shown that in the representation $\exp y = H|K$, $h \in H \Rightarrow h \ll \exp y$ and $k \in K \Rightarrow \exp y \ll k$.

We now look at the representation $\exp z = F|G$. Suppose $u < z$. Then u contributes terms such as $\exp u[z-u]_n$ to F and $\frac{\exp u}{[u-z]_{2n+1}}$ to G. If $z-u$ is not infinitesimal, then neither $[z-u]_{n+1} - [z-u]_n$ nor $\frac{1}{[u-z]_{2n+1}} - \frac{1}{[u-z]_{2n+3}}$ is infinitesimal. Also, $\exp(z-u) - [z-u]_n \geq [z-u]_{n+1} - [z-u]_n$ hence $\exp[z-u] - [z-u]_n$ is not infinitesimal. Similarly $\frac{1}{[u-z]_{2n+1}} - \exp[z-u]$ is not infinitesimal. Since all the expressions considered are finite, we obtain by multiplication with exp u that

$\exp u \sim \exp u[z-u]_{n+1} - \exp u[z-u]_n \sim \exp u[\exp(z-u)] - \exp u[z-u]_n \sim$
$\frac{\exp u}{[u-z]_{2n+1}} - \frac{\exp u}{[u-z]_{2n+3}} \sim \frac{\exp u}{[u-z]_{2n+1}} - \frac{\exp u}{\exp[u-z]}$. Note of course that

$\exp u \exp(z-u) = \frac{\exp u}{\exp(u-z)} = \exp z$. What the above essentially says is that the various orders of magnitude of differences of elements of F and G with each other and with exp z are the same.

If $z-u$ is infinitesimal we need more caution. Suppose $z-u \sim \omega^a$, i.e. a is the first exponent occurring in the normal form of $z-u$. Then $[z-u]_{n+1} - [z-u]_n \sim \omega^{(n+1)a}$. By formal multiplicaion

$\exp[z-u] - (z-u]_n \sim \omega^{(n+1)a}$. Again using formal reciprocals we see that $\frac{1}{[u-z]_{2n+1}} - \frac{1}{[u-z]_{2n+3}} \sim \omega^{(2n+2)a}$ and $\frac{1}{[u-z]_{2n+1}} - (\exp(z-u) \sim \omega^{(2n+2)a}$.

Hence $\exp u[z-u]_{n+1} - \exp u[z-u]_n \sim \exp z - \exp u[z-u]_n$

$\frac{\exp u}{[u-z]_{2n+1}} - \frac{\exp u}{[u-z]_{2n+3}} \sim \frac{\exp u}{[u-z]_{2n+1}} - \exp z$ and

$\exp z - \exp u[z-u]_{2n+1} \sim \frac{\exp u}{[u-z]_{2n+1}} \exp z$.

We have similar results if $u > z$.

Now we are ready to study the terms in the product $\exp y \exp z = (H|K)(F|G)$. Recall that this has the form $\{h \exp z + f \exp y - hf,\ k \exp z + g \exp y - kg\}|\{h \exp z + g \exp y - hg,\ k \exp z + f \exp y - kf\}$ where $h \in H$, $k \in H$, $f \in F$, and $g \in G$. Recall also that we desire to prove the cofinality of terms of the form $f \exp y$ and $g \exp y$. First, consider $h \exp z + f \exp y - hf = f \exp y + h(\exp z-f)$. f has a form such as $\exp u[z-u]_n$. Suppose $f' = \exp u[z-u]_{n+1}$. Then $f' - f \sim \exp z - f$. Also $h \ll \exp y$. Therefore $f \exp y + h(\exp z-f) < f \exp y + \exp y(f'-f) = f' \exp y$ which gives us precisely what we need. Now consider $k \exp z + g \exp y - kg = g \exp y - k(g-\exp z)$. g has a form such as $\frac{\exp u}{[u-z]_{2n+1}}$. If we let $f = \exp u[z-u]_{2n+1}$ then we have $g - \exp z \sim \exp z - f$. $g - f \sim g-\exp z$. Also $\exp y \ll k$. Therefore $g \exp y - k(g-\exp z) < g \exp y - \exp y(g-f) = f \exp y$.

The upper terms in the representation of the product can be handled the same way. For this purpose express $h \exp z + g \exp y - hg$ in the form $g \exp y - h(g-\exp z)$ and $k \exp z + f \exp y - kf$ in the form $f \exp y + k(\exp z-f)$.

This completes the proof.

We are now ready to study the purely infinite case.

B SPECIALIZATION TO PURELY INFINITE NUMBERS

We begin by inviting the reader to share with us the pleasure of the simplification obtained in the case of purely infinite numbers.

Recall that a purely infinite number is a number in which all the exponents in its normal form are positive. The number itself may be positive or negative. Also 0 is a purely infinite number! This may sound strange at first but it is consistent with the usual meaning in logic of universal quantification over the empty set and is natural for the theoretical development. Such numbers can be characterized also in a way which does not mention the normal form. Just as surreal numbers of the form ω^a are the elements of smallest length among numbers of the same multiplicative order of magnitude, purely infinite numbers are elements of smallest length among numbers of the same additive order of magnitude. Although this point of view does not play a significant role in our work it does add a certain beauty to some later theorems such as theorem 10.7.

If x is a purely infinite number then we already know that every initial segment y satisfies $|x-y|$ is infinite. Hence no terms of the form $[z]_{2n+1}$ where z is negative arise in the definition of $\exp x$. Thus lower initial segments of x do not contribute to upper terms in the definition of $\exp x$ and vice versa. Therefore we have $\exp = \{0, \exp x'[x-x']_n\} | \{\frac{\exp x''}{[x''-x]_n}\}$. Also for positive infinite z $z^{n+1} > [z]_n$ and $[z]_{n+1} > z_n$. Hence by mutual cofinality we may simplify the representation of $\exp x$ to $\{0, \exp x'(x-x')^n\} | \{\frac{\exp x''}{(x''-x)^n}\}$. Thus we no longer have to bother with sums of the form $[x]_n$.

We now go one step further by showing that we need consider purely infinite initial segments only. In fact, suppose $x' = y+z$ where x' is a lower initial segment of x, y is purely infinite, and z is finite. Since $x-x'$ is infinite it follows that $y < x$. Also y is an initial segment of x' hence also an initial segment of x. We claim now that terms in the representation of $\exp x$ contributed by x' are mutually cofinal with terms contributed by y. First $\exp x' = \exp(y+z) = \exp y \exp z \sim \exp y$ since $\exp z$ is finite. Now consider an arbitrary expression of the form $(u+v)^n$ where u is positive infinite and v is finite.

If v is negative then certainly $(u+v)^n \leq u^n \ll u^{n+1}$. If v is positive we have:

$$(u+v)^n = \sum_{i=0}^{n} \binom{n}{i} u^i v^{n-i} \leq \sum_{i=0}^{n} \binom{n}{i} u^n v^{n-i} = u^n \left[\sum_{i=0}^{n} \binom{n}{i} v^{n-i} \right] \ll u^{n+1}.$$

So in either case $(u+v)^n \ll u^{n+1}$. If we apply this to the above we obtain $\exp x'(x-x')^n \sim \exp y(x-x')^n = \exp y[x-(y+z)]^n = \exp y(x-y-z)^n \ll \exp y(x-y)^{n+1}$. Hence $\exp x'(x-x')^n < \exp y(x-y)^{n+1}$. A similar argument applies to upper initial segments. We state the final simplification as theorem 10.5.

Theorem 10.5. For purely infinite x $\exp x$ has the form $\{0, \exp x'(x-x')^n\} | \{\frac{\exp x''}{(x''-x)^n}\}$ where x' runs through all purely infinite lower initial segments of x and x'' runs through all purely infinite upper initial segments of x.

As an example consider $\exp \omega$. 0 is the only purely infinite lower segment. Therefore $\exp \omega = \{0, \omega^n\} | \phi = \omega^\omega$. Since ω is, of course, different from e this illustrates that ω^x as defined in chapter five should not be regarded as exponentiation in the sense that is being developed here. We shall see later what arises naturally as the value for $\exp (x \ln \omega)$ after we study the function $\ln$.

A similar result applies if x is expressed in a form $F|G$ which is not necessarily canonical but some caution is required. Suppose that $y \in F \cup G \Rightarrow |x-y|$ is infinite. In general the purely infinite part of an element in $F \cup G$ need not be in $F \cup G$. [By the infinite part of an element y is meant the unique purely infinite number z such that $y-z$ is finite.] However, by the uniformity theorem for $\exp x$ we can start with using F and G and still do the same simplification as in the canonical representation but instead of taking subsets of $F \cup G$ at the end we use the infinite parts of all the elements of F for the lower terms and of G for the upper terms.

In future x, y, and z will refer to purely infinite numbers unless specifically stated otherwise. For reference we state two inequalities which are similar to ones we had earlier. Assume x and y are positive;

$$(x+y)^{2n} \geq x^n y^n$$

$$x^{n+1} y^{n+1} \geq (x+y)^n.$$

The first is clear since by the binomial theorem $\binom{2n}{n} x^n y^n$ is one of the terms in the expansion of $(x+y)^{2n}$. The second is clear since a typical term $\binom{n}{i} x^i y^{n-i}$ in the expansion of $(x+y)^n$ satisfies

$\binom{n}{i}x^i y^{n-i} << x^{n+1}y^{n+1}$. Although the latter inequality may be strengthened there is not much point to it for our purpose. (It is much like showing that one can use $\frac{\varepsilon}{32}$ instead of $\frac{\varepsilon}{64}$ in analysis!)

Recall that by theorem 10.1, if $x < y$, then $\exp x[y-x]_n < \exp y$. Hence $\exp x(y-x)^n < \exp x[y-x]_{n+1} < \exp y$. Thus we may replace the term $[y-x]_n$ by $(y-x)^n$ in the inequality stated as part of the above theorem. Since $y-x$ is infinite, this shows immediately that $x < y \to \exp x << \exp y$. As we recall, the function ω^x has this property, but there is a crucial distinction, namely for ω^x this applies to the whole domain, but to $\exp x$ this applies only to the subclass of purely infinite numbers. In fact, if $y-x$ is finite then $e^x \sim e^y$. At any rate, because of the above property the proof of the addition formula for $\exp x$ resembles the proof of the corresponding formula for ω^x.

Theorem 10.6. $\exp(x+y) = \exp x \exp y$.

Proof. We use induction. Without loss of generality a typical element in the representation of $x+y$ has the form $x^0 + y$. Since x is purely infinite, $|x-x^0|$ is infinite, hence so is $|(x+y)-(x^0+y)|$. Therefore in the computation of $\exp(x+y)$ we may use the infinite parts of elements such as x^0+y. Since y is purely infinite the infinite part of x^0+y is y plus the infinite part of x^0. Also, as x^0 runs through all proper initial segments of x the infinite parts of x^0 run through all proper purely infinite initial segments and the infinite part of x^0 is on the same side of x as x^0. This permits us to write $\exp(x+y)$ in the form $\{0, \exp(x'+y)[(x+y)-(x'+y)]^n, \exp(x+y')[(x+y)-(x+y')]^n\} | \{\frac{\exp(x''+y)}{[(x''+y)-(x+y)]^n}, \frac{\exp(x+y'')}{[(x+y'')-(x+y)]^n}\}$ where $\exp x$ has the form $\{0, \exp x[x-x']^n\} | \{\frac{\exp x''}{(x''-x)^n}\}$ and similarly for y. Using the inductive hypothesis, a typical lower term simplifies to $\exp x' \exp y(x-x')^n$ and an upper term to $\frac{\exp x'' \exp y}{(x''-x)^n}$. 0 is, of course, also a lower term. These are terms occurring in the representation of $\exp x \exp y$ since 0 is a lower element of both x and y. Thus, as in the proof of theorem 10.4, it remains to prove that these terms are cofinal with respect to the other terms in the formula for the product. If x or y is 0, there are no other terms so the result is clear. (That case is trivial anyway and could just as well be eliminated in advance.)

First, we have a lower term such as $\exp x \exp y'(y-y')^m + \exp y \exp x'(x-x')^n - \exp y'(y-y')^m \exp x'(x-x')^n$. Suppose $\exp x \exp y'(y-y')^m \leq \exp y \exp x'(x-x')^n$. Then
$\exp x \exp y'(y-y')^m + \exp y \exp x'(x-x')^n - \exp y'(y-y')^m \exp x'(x-x')^n$
$\leq \exp x \exp y'(y-y')^m + \exp y \exp x'(x-x')^n \leq 2 \exp y \exp x'(x-x')^n$
$\leq \exp y \exp x'(x-x')^{n+1}$. A similar result applies if
$\exp y \exp x'(x-x')^n \leq \exp x \exp y'(y-y')^m$.

Now consider a lower term such as

$\dfrac{\exp x \exp y''}{(y''-y)^m} + \dfrac{\exp y \exp x''}{(x''-x)^n} - \dfrac{\exp y''}{(y''-y)^m}\dfrac{\exp x''}{(x''-x)^n}$. We know that

$\dfrac{\exp x''}{(x''-x)^n} \gg \dfrac{\exp x''}{(x''-x)^{n+1}} > \exp x$ hence $\dfrac{\exp y''}{(y''-y)^m}\dfrac{\exp x''}{(x''-x)^n} \gg \dfrac{\exp y''}{(y''-y)^m} \exp x$.

Similarly $\dfrac{\exp y''}{(y''-y)^m}\dfrac{\exp x''}{(x''-x)^n} \gg \dfrac{\exp x''}{(x''-x)^n} \exp y$.

Therefore the above expression is negative, so that for cofinality it suffices to take the term zero.

Finally we must consider a term such as

$\exp x \exp y'(y-y')^m + \exp y \dfrac{\exp x''}{(x''-x)^n} - \exp y'(y-y')^m \dfrac{\exp x''}{(x''-x)^n}$. Now

$\dfrac{\exp y \exp x''}{(x''-x)^n} \gg \exp y'(y-y')m \dfrac{\exp x''}{(x''-x)^n}$. Therefore certainly

$\exp x \exp y'(y-y')m + \exp y \dfrac{\exp x''}{(x''-x)^n} - \exp y'(y-y')m \dfrac{\exp x''}{(x''-x)^n} \geq$

$\exp y \dfrac{\exp x''}{(x''-x)^n} - \exp y'(y-y')m \dfrac{\exp x''}{(x''-x)^n} \geq \frac{1}{2} \exp y \dfrac{\exp x''}{(x''-x)^n} \geq \dfrac{\exp y \exp x''}{(x''-x)^{n+1}}$.

A similar result applies if we interchange the role of x and y. This proves the cofinality and completes the proof.

<u>Corollary 10.1</u>. $\exp(x+y) = \exp x \exp y$ for <u>all</u> x and y.

<u>Proof</u>. This follows immediately from theorem 10.6 with the help of theorem 10.4 and the corresponding result for finite x and y.

Corollary 10.2. $\exp x \exp(-x) = 1$.

Our next aim is to show that as x runs through all purely infinite numbers, $\exp x$ runs through all numbers of the form ω^a where $a \geq 0$. (Of course, if $a = 0$ we obtain $\exp 0 = 1 = \omega^0$ so our main concern is with $a > 0$.) One direction is almost immediate. The other direction will be proved by means of the function $\ln x$.

Theorem 10.7. $\exp x$ is a power of ω.

Proof. We already know that every lower term such as $\exp x'(x-x)^n$ satisfies $\exp x'(x-x')^n \ll \exp x$ and similarly every upper term such as $\frac{\exp x''}{(x''-x)^n}$ satisfies $\exp x \ll \frac{\exp x''}{(x''-x)^n}$. Hence every number y of the same order of magnitude as x also satisfies $\exp x'(x-x')^n < y < \frac{\exp x''}{(x''-x)^n}$. Therefore x is an initial segment of y. So by theorem 5.3 x is a power of ω.

The above result is especially striking if one recalls that purely infinite elements are canonical elements in an additive order of magnitude. The theorem says that such elements are mapped into canonical elements in a multiplicative order of magnitude. This is further heuristic evidence that exp behaves the way an exponential should.

We now define $\ln x$. This definition will be made only for x of the form ω^b.

Definition. $\ln(\omega^b) = \{\ln(\omega^{b'})+n,\ \ln(\omega^{b''}) - \omega^{\frac{b''-b}{n}}\} | \{\ln(\omega^{b''})-n,\ \ln(\omega^{b'}) + \omega^{\frac{b-b'}{n}}\}$ where n runs through all positive integers.

Note that lower initial segments in b contribute to upper elements in the representation of $\ln(\omega^b)$. This is essential. For example, let $b = 1$. Then $\ln \omega = \{\ln(\omega^0)\}+n | \{\ln(\omega^0)+(\omega^1)^{\frac{1}{n}}\}$. With no upper terms this would be $\{n\}|\phi = \omega$. Since $\exp \omega = \omega^\omega$ with this definition $\ln x$ would not be the inverse function of $\exp x$. On the other hand the given definition leads to $\ln \omega = \{n\}|\{\omega^{\frac{1}{n}}\} = \omega^{\frac{1}{\omega}}$. This does seem more reasonable a priori. In fact, let us compute $\exp(\omega^{\frac{1}{\omega}})$.

$\frac{1}{\omega} = \{0\}|\{\frac{1}{n}\}$. Then $\omega^{\frac{1}{\omega}} = \{0,n\}|\{\omega^{\frac{1}{n}}\}$. Hence by restricting ourselves to infinite parts $\exp(\omega^{\frac{1}{\omega}}) = \{(\exp 0)\omega^{\frac{n}{\omega}}\}|\{\frac{\exp(\omega^{\frac{1}{n}})}{(\omega^{\frac{1}{n}}-\omega^{\frac{1}{\omega}})^m}\}$. The lower terms are simply $\omega^{\frac{n}{\omega}}$. Now by the addition formula $\exp(\omega^{\frac{1}{n}}) = (\exp \omega)^{\frac{1}{n}} = (\omega^{\omega})^{\frac{1}{n}} = \omega^{\frac{\omega}{n}}$. Also $(\omega^{\frac{1}{n}}-\omega^{\frac{1}{\omega}})^m \le (\omega^{\frac{1}{n}})^m = \omega^r$ for some real number r (in fact $\frac{m}{n}$). So $\frac{\exp(\omega^{\frac{1}{n}})}{(\omega^{\frac{1}{n}}-\omega^{\frac{1}{\omega}})^m} \ge \omega^{\frac{\omega}{n}-r} > \omega$. Now $\omega = \{n\}|\phi$. Since $\omega^{\frac{n}{\omega}} < \omega$ and $\omega^{\frac{n}{\omega}}$ is infinite if $n \ge 1$, the conditions of the cofinality theorem are satisfied and $\exp(\omega^{\frac{1}{\omega}}) = \omega$.

Incidentally, the reader may be interested in experimenting with various other computations using the definition. We prefer to postpone a discussion of explicit results until we have proved some remarkable results which will simplify the computation tremendously. We still have a distance to go before this. For example, we don't yet even know that $\ln x$ is defined!

Theorem 10.8. $\ln x$ is defined. Furthermore, if $a > b$ then $\ln(\omega^a) - \ln(\omega^b)$ is positive infinite and $\ln(\omega^a) - \ln(\omega^b) < \omega^{\frac{a-b}{n}}$ for all positive integers n. In particular, if $a > 0$ then $\ln(\omega^a)$ is positive infinite and $\ln(\omega^a) < \omega^{\frac{a}{n}}$.

Proof. It is no longer surprising that we use induction. We now show that the lower terms are really less than the upper terms. Since $\ln(\omega^{b''}) - \ln(\omega^{b'})$ is infinite, then certainly $\ln(\omega^{b'})+n < \ln(\omega^{b''})-m$. Since $\omega^{\frac{b-b'}{n}}$ is infinite then certainly $\ln(\omega^{b'})+n < \ln(\omega^{b'}) + \omega^{\frac{b-b'}{m}}$. However, such an inequality is not adequate for our purpose since we must consider two arbitrary initial segments of b not necessarily alike. (The use of the same b', which is a natural mistake, leads to an incomplete proof.) So we must show that $\ln(\omega^c)+n < \ln(\omega^d) + \omega^{\frac{b-d}{m}}$

where c and d are two lower initial segments of b. This is trivial if $d \geq c$. If $c > d$ we have $\ell n(\omega^c) - \ell(\omega^d) < \omega^{\frac{c-d}{m}} < \omega^{\frac{b-d}{m}} - n$. Hence $\ell n(\omega^c)+n < \ell n(\omega^d) + \omega^{\frac{b-d}{m}}$. Similarly for upper initial segments c and d of b we must show that $\ell n(\omega^c) - \omega^{\frac{c-b}{m}} < \ell n(\omega^c)-n$. As before this is trivial if $c \leq d$. If $c > d$ we have $\ell n(\omega^c) - \ell n(\omega^d) < \omega^{\frac{c-d}{m}} < \omega^{\frac{c-b}{m}} -n$ hence $\ell n(\omega^c) - \omega^{\frac{c-b}{m}} < \ell n(\omega^d)-n$.

It remains to prove that $\ell n(\omega^{b''}) - \omega^{\frac{b''-b}{n}} < \ell n(\omega^{b'}) + \omega^{\frac{b-b'}{m}}$.

By the inductive hypothesis we have $\ell n(\omega^{b''}) - \ell n(\omega^{b'}) < \omega^{\frac{b''-b'}{n}}$ for all n. Now $\omega^{\frac{b''-b}{2n}} = [\omega^{\frac{b''-b}{n}} \omega^{\frac{b-b'}{n}}]^{\frac{1}{2}} \leq \frac{1}{2}[\omega^{\frac{b''-b}{n}} + \omega^{\frac{b-b'}{n}}]$ by the arithmetico-geometric inequality which is valid in any ordered field. Hence $\omega^{\frac{b''-b}{2n}} \leq \omega^{\frac{b''-b}{n}} + \omega^{\frac{b-b'}{n}}$. Since $\omega^{\frac{x}{n}}$ is a decreasing function of n then for $r = \max(m,n)$ we have $\omega^{\frac{b''-b}{2r}} \leq \omega^{\frac{b''-b}{n}} + \omega^{\frac{b-b'}{m}}$. Hence $\ell n(\omega^{b''}) - \ell n(\omega^{b'}) < \omega^{\frac{b''-b}{n}} + \omega^{\frac{b-b'}{m}}$ so $\ell n(\omega^{b''}) - \omega^{\frac{b''-b}{n}} < \ell n(\omega^{b'}) + \omega^{\frac{b-b'}{m}}$.

We now know that $\ell n\ (\omega^b)$ is defined. For $a > b$ it is immediate from the definition that $\ell n(\omega^a) - \ell n(\omega^b)$ is infinite if one of a and b is an initial segment of the other. In the general case it suffices to consider the common initial segment of x and y since the sum of two positive infinite numbers is clearly infinite.

For the final inequality let c be the common initial segment of a and b where $a < b$ so that $a < c < b$. (Again if $c = a$ or b the result is trivial.) Then from the definition we have $\ell n(\omega^a) > \ell n(\omega^c) - \omega^{\frac{c-a}{n}}$ and $\ell n(\omega^b) < \ell n(\omega^c) + \omega^{\frac{b-c}{n}}$. Then $\ell n(\omega^b)-\ell n(\omega^a) < \omega^{\frac{b-c}{n}} + \omega^{\frac{c-a}{n}}$. Now $\omega^{\frac{b-c}{n}} \ll \omega^{\frac{b-a}{n}}$ and $\omega^{\frac{c-a}{n}} \ll \omega^{\frac{b-a}{n}}$.

Therefore $\omega^{\frac{b-c}{n}} + \omega^{\frac{c-a}{n}} < \omega^{\frac{b-a}{n}}$. This finally gives us the inequality $\ell n(\omega^b) - \ell n(\omega^a) < \omega^{\frac{b-a}{n}}$. This completes the proof.

We would like to show next that $\ell n(x)$ is the inverse function of $\exp x$ where the domain of exp is restricted to the purely infinite numbers. For this purpose we first need the following theorem.

Theorem 10.8. $\ell n(\omega^b)$ is purely infinite for all b.

Proof. Let $F|G$ be the representation of $\ell n(\omega^b)$ given by the definition. We claim that $c \in F \Rightarrow (\exists d \in F)(d \geq c+1)$ and $c \in G \Rightarrow (\exists d \in G)(d \leq c-1)$. This is easy to see. If c has the form $\ell n(\omega^{b'})+n$ we get $d = \ell n(\omega^{b'}) + (n+1)$ and if c has the form $\ell n(\omega^{b''}) - \omega^{\frac{b''-b}{n}}$ then we may choose $d = \ell n(\omega^{b''}) - \omega^{\frac{b''-b}{n+1}}$. (In the second case $d-c$ is infinite.) A similar argument applies to upper terms. By induction 1 can be replaced by any positive integer.

This is enough to show that $\ell n(\omega^b)+r$ satisfies $F < \ell n(\omega^b)+r < G$ for any real r. Hence $\ell n(\omega^b)$ is an initial segment of its purely infinite part, so $\ell n(\omega^b)$ must itself be purely infinite.

We are now ready for the theorem which completes the basic theory.

Theorem 10.9. $\exp\ \ell n(\omega^b) = \omega^b$.

Proof. Again we use induction.

$$\ell n(\omega^b) = F|G \text{ where } F = \{\ell n(\omega^{b'})+n,\ \ell n(\omega^{b''}) - \omega^{\frac{b''-b}{n}}\} \text{ and}$$

$G = \{\ell n(\omega^{b''})-n,\ \ell n(\omega^{b'}) + \omega^{\frac{b-b'}{n}}\}$. We already know that $c \in F \Rightarrow (\exists d \in F)(d \geq c+1)$ and $c \in G \Rightarrow (\exists d \in G)(d \leq c-1)$. This implies that $c \in F \Rightarrow \ell n(\omega^b)-c$ is infinite and $d \in G \Rightarrow d-\ell n(\omega^b)$ is infinite. Thus by our earlier remarks we may use the infinite parts of the elements in F and G to compute $\exp \ell n(\omega^b)$. By theorem 10.8 and the fact that any positive power of ω is infinite, the infinite parts of the elements in F have the forms $\ell n(\omega^{b'})$ and $\ell n(\omega^{b''}) - \omega^{\frac{b''-b}{n}}$, and similarly those in G have the form $\ell n(\omega^{b''})$ and $\ell n(\omega^{b'}) + \omega^{\frac{b-b'}{n}}$. Therefore the typical lower terms besides 0 in the formula for $\exp[\ell n(\omega^b)]$ have the form $\exp[\ell n(\omega^{b'})][\ell n(\omega^b)-\ell n(\omega^{b'})]^n$ and

$\exp[\ell n(\omega^{b''})-\omega^{\frac{b''-b}{n}}]\{\ell n(\omega^b)-[\ell n(\omega^{b''})-\omega^{\frac{b''-b}{n}}]\}^m$, and similarly the upper terms have the form $\dfrac{\exp(\ell n(\omega^{b''})}{[\ell n(\omega^{b''})-\ell n(\omega^b)]^n}$ and $\dfrac{\exp[\ell n(\omega^{b'})+\omega^{\frac{b-b'}{n}}]}{[\ell n(\omega^{b'})+\omega^{\frac{b-b'}{n}} - \ell n(\omega^b)]^m}$.

We use cofinality to show that this gives rise to ω^b which is $\{0,\omega^b r\}|\{\omega^{b''}s\}$.

First, we check the betweenness condition.

By the inductive hypothesis $\exp\ \ell n(\omega^{b'})[\ell n(\omega^b)-\ell n(\omega^{b'})]^n$

$= \omega^{b'}[\ell n(\omega^b)-\ell n(\omega^{b'})]^n < \omega^{b'}(\omega^{\frac{b-b'}{n}})^n = \omega^{b'}\omega^{b-b'} = \omega^b$. Similarly

$\dfrac{\exp[\ell n(\omega^{b''})]}{[\ell n(\omega^{b''})-\ell n(\omega^b)]} = \dfrac{\omega^{b''}}{[\ell n(\omega^{b''})-\ell n(\omega^b)]^n} > \dfrac{\omega^{b''}}{\omega^{b''-b}} = \omega^b$. Fortunately the other terms are not as hard to deal with as it may appear. First,

$$\exp[(\ell n(\omega^{b''})-\omega^{\frac{b''-b}{n}}]\{\ell n(\omega^b) - [\ell n(\omega^{b''})- \omega^{\frac{b''-b}{n}}]\}^m$$

$$= \frac{\exp[\ell n(\omega^{b''})]}{\exp(\omega^{\frac{b''-b}{n}})}\{\ell n(\omega^b) - [\ell n(\omega^{b''}) - \omega^{\frac{b''-b}{n}}]\}^m$$

$$< \frac{\omega^{b''}}{\exp(\omega^{\frac{b''-b}{n}})}(\omega^{\frac{b''-b}{n}})^m \quad \text{since} \quad \ell n(\omega^{b''}) > \ell n(\omega^b)$$

$$< \frac{b''}{(\omega^{\frac{b''-b}{n}})^{n+m}}(\omega^{\frac{b''-b}{n}})^m = \frac{\omega^{b''}}{\omega^{b''-b}} = \omega^b. \quad \text{Similarly}$$

$$\frac{\exp\left[\ell n(\omega^{b'})+\omega^{\frac{b-b'}{n}}\right]}{\left[\left(\ell n(\omega^{b'})+\omega^{\frac{b-b'}{n}}\right)-\ell n(\omega^{b})\right]^m} = \frac{\omega^{b'}\exp\left(\omega^{\frac{b-b'}{n}}\right)}{\left[\ell n(\omega^{b'})+\omega^{\frac{b-b'}{n}}-\ell n(\omega^{b})\right]^m} > \frac{\omega^{b'}\exp\left(\omega^{\frac{b-b'}{n}}\right)}{\left(\omega^{\frac{b-b'}{n}}\right)^m} >$$

$$\frac{\omega^{b'}\left(\omega^{\frac{b-b'}{n}}\right)^{m+n}}{\left(\omega^{\frac{b-b'}{n}}\right)^m} = \omega^{b'}\omega^{b-b'} = \omega^{b}.$$ This verifies the betweenness condition.

The cofinality condition is easier to check since the simpler lower and upper terms are all we need for the purpose.

$\omega^{b'}r$ is a typical lower term in the representation of ω^b. One of the lower terms found in the representation of $\ell n(\omega^b)$ is $\exp[\ell n(\omega^{b'})][\ell n(\omega^b-\ell n(\omega^{b'})]$. The first factor is $\omega^{b'}$. $\ell n(\omega^b) > \ell n(\omega^{b'})$ and both are purely infinite. Hence $\ell n(\omega^b) - \ell n(\omega^{b'})$ is infinite. So we may take $n = 1$ and we have an element which is larger than $\omega^{b'}r$ since r is, of course, finite. Similarly $\omega^{b''}s$ is a typical upper term in the representation of ω^b. Here we consider $\dfrac{\exp[\ell n(\omega^{b''})]}{[\ell n(\omega^{b''})-\ell n(\omega^{b'})]^n}$ which is $\omega^{b''}\varepsilon$ where ε is infinitesimal. (Again we may use $n = 1$.) This is smaller than $\omega^{b''}s$. The proof is now complete.

It is immediate from theorem 10.9 that the function $\exp x$ defined for x purely infinite is onto the class of numbers of the form ω^a. By theorem 10.4 and earlier information on $\exp x$ for finite x it follows that the range of $\exp x$ where x runs through the class of all surreal numbers consists of all numbers of the form $\omega^a b$ where b is the class of all positive finite non-infinitesimal numbers. This is precisely the class of all numbers of the form $\sum_{i<\alpha} \omega^{a_i} b_i$ with $b_0 > 0$ since the latter can be expressed in the form $\omega^{a_0}(b_0 + \sum_{1 \leq i<\alpha} \omega^{c_i} b_i)$. This is in turn the class of all positive surreals. Thus we have

<u>Corollary 10.3</u>. $\exp x$ is <u>onto</u> the class of all positive numbers.

We now know that $\exp x$ behaves as it should with respect to the main properties expected of an exponential function. Thus we have good philosophical grounds in designating our function as "the exponential function." Squeamishness about proper classes can be handled by the techniques of chapter six.

With the basics out of the way we are now ready to move towards the exotic surprises of the exponential function.

C REDUCTION TO THE FUNCTION g

The central problem of interest can be expressed as follows. Given $x = \sum_{i<\alpha} \omega^{a_i} r_i$, what can we say about $\exp x$? I.e. we desire to be as explicit as possible. We repeat that our interest is limited to the purely infinite case since we already know what goes on in the finite case. It turns out there is much that can be said but the subject is non-trivial, i.e. there is no single theorem which closes the subject as rapidly as in the cases where x is real or infinitesimal. A tremendous amount of further simplification is possible; however, in spite of this there are enough complications remaining so that the subject remains substantial. (This means that we don't go out of business by exhausting the subject too rapidly.)

We are finally ready for the first of the beautiful results.

Theorem 10.10. If $a > 0$ then $\exp(\omega^a)$ has the form ω^{ω^b}.

Proof. We do not use induction!! Since we are studying purely infinite numbers only at this time, the condition $a > 0$ is understood. Hence in the canonical representation of a all terms are non-negative. Thus if we express ω^a as $\{0,\omega^{a'}r\}|\{\omega^{a''}s\}$ all terms are purely infinite and have infinite distance from ω^a. Therefore we may use these terms for the computation of $\exp(\omega^a)$. So $\exp(\omega^a)$ is
$\{0,(\exp 0)(\omega^a)^n,[\exp(\omega^{a'}r)](\omega^a-\omega^{a'}r)^n\}|\{\frac{\exp(\omega^{a''}s)}{(\omega^{a''}s-\omega^a)^n}\}$. We now use mutual cofinality to simplify. The term 0 is superfluous because of the second group of lower terms which are of the form ω^{na}. Now
$(\omega^a-\omega^{a'}r)^{n+1} > (\omega^a)n > (\omega^a-\omega^{a'}r)^n$ and
$(\omega^{a''}s-\omega^a)^{n+1} > (\omega^{a''})n > (\omega^{a''}s-\omega^a)^n$. (We may just as well assume that $s \leq 1$ by cofinality or alternatively we can use the inequality $(\omega^{a''})^{n+1} > (\omega^{a''}s-\omega^a)n$.) At any rate, the representation for $\exp(\omega^a)$

simplifies to $\{(\omega^{na},[\exp(\omega^{a'}r)]\omega^{na}\}|\{\frac{\exp(\omega^{a''}s)}{\omega^{na''}}\}$. Since 0 is a possible value of a' and 1 a possible value of r the first group of lower terms are superfluous. Since in general $\exp x > \exp O(x^n) = x^n$ we have $\exp(\omega^{a''}\frac{s}{2}) > (\omega^{a''}\frac{s}{2})^{n+1} > \omega^{na''}$. Thus
$\exp(\omega^{a''}\frac{s}{2}) = \frac{\exp(\omega^{a''}s)}{\exp(\omega^{a''}\frac{s}{2})} < \frac{\exp(\omega^{a''}s)}{\omega^{na''}}$. Hence the upper terms may be limited to terms of the form $\exp(\omega^{a''}s)$. So $\exp(\omega^{a})$ simplifies further to $\{[\exp(\omega^{a'}r)]\omega^{na}\}|\{\exp(\omega^{a''}s)\}$. Also we may limit r to integers and s to dyadic fractions. Hence we may write $\exp(\omega^{a'}r)$ as $[\exp(\omega^{a'})]^r$ and $\exp(\omega^{a''}s)$ as $[\exp(\omega^{a''})]^s$.

By theorem 10.7 $\exp(\omega^{a'})$ and $\exp(\omega^{a''})$ are powers of ω. (Note that we are not assuming the stronger fact that $\exp(\omega^{a'})$ and $\exp(\omega^{a''})$ are iterated powers of ω as we would in a typical proof by induction.) Let $\exp(\omega^{a'}) = \omega^{b'}$ and $\exp(\omega^{a''}) = \omega^{b''}$. Then $\exp(\omega^{a})$ can be expressed as

$$\{\omega^{b'r}\omega^{na}\}|\{(\omega^{b''s}\}.$$

Note that we use the additivity theorem for both $\exp x$ and ω^x to justify the above. We may just as well write this as

$$\{0,\omega^{b'r}\omega^{na}\}|\{\omega^{b''s}\} = \{0,\omega^{b'r+na}\}|\{\omega^{b''s}\}.$$

Since the set $\{b'r+na\}$ has no maximum and $b''s$ has no minimum, this may be expressed as a power of ω, specifically ω^b where $b = \{b'r+na\}|\{b''s\}$. It is redundant to try to show that $b'r+na < b''s$ since we began with something which was defined and used mutual cofinality at every step, so we already know that $\omega^{b'r+na} < \omega^{b''s}$. Consider the representation $b = \{0,b'r+na\}|\{b''s\}$. If $b \sim c$ then $b'r+na < c < b''s$ hence b is an initial segment of c. Therefore b is a power of ω. Let $b = \omega^c$. Then $a = \omega^{\omega^c}$.

By theorem 10.10 $\exp x$ induces a function g with domain consisting of all positive surreals such that $\exp(\omega^a) = \omega^{\omega^{g(a)}}$. We now use the above proof to obtain an inductive formula for $g(a)$. For this purpose we define ind a as follows: If a has the normal form $\sum_{i<\alpha} \omega^{b_i}r_i$ then ind $a = b_0$. b_0 can also be characterized as the unique surreal x such that $a \sim \omega^x$. We also use the temporary notation $\exp(\omega^x) = \omega^{f(x)}$ for $x > 0$ so that $f(a) = \omega^{g(a)}$. If $x = 0$ we set $f(0) = 0$. From the above proof we have using this notation $f(a) = \{rf(a')+na\}|\{sf(a'')\}$. Since one value of a' is 0 in which case $f(a') = 0$, we may certainly write $f(a) = \{0,rf(a')+na\}|\{sf(a'')\}$ by cofinality. Now among all the elements of the form a' we single out 0 and write this as

$$f(a) = \{0,na,rf(a')+na\}|\{sf(a'')\}.$$

Of course, there may be no other a' just as there may be no a'' altogether.

We may now replace $f(x)$ by $\omega^{g(x)}$ which is valid for $x > 0$. So we obtain $f(a) = \{0,na,r\omega^{g(a')}+na\}|\{s\omega^{g(a'')}\}$. Let $c = \text{ind } a$, i.e. $a \sim \omega^c$. Then na is equicofinal with $n\omega^c$ and $r\omega^{g(a')}+na$ is equicofinal with $r\omega^{g(a')} + n\omega^c$ which is in turn equicofinal with $n\omega^{\max[g(a'),c]}$. Hence

$$f(a) = \{0,n\omega^c,n\omega^{\max[g(a'),c]}\}|\{s\omega^{g(a'')}\}.$$

Since $f(a) = \omega^{g(a)}$ we can now determine $g(a)$. In fact, $g(a) = \{c,\max[g(a'),c]\}|\{g(a'')\}$ which is the same as

$$\{c,g(a')\}|\{g(a'')\} \text{ by cofinality.}$$

We have thus ended up with a rather remarkably simple-looking formula. However, the reader is warned that the presence of the term c gives rise to tricky phenomena. For example, the presence of c prevents g from simply being the identity function as we shall see. In fact, g can take on negative values as we shall also see. This is consistent with what we have since ω^{ω^x} is <u>positive infinite</u> for <u>all</u> x. Note that there is no zero among the lower terms in the expression for $g(a)$. This illustrates that in general the inclusion of 0 among the lower terms must be treated with caution to ensure that it is legal.

In view of its importance for the rest of our work we state the formula for $g(x)$ as a theorem.

<u>Theorem 10.11</u>. If a is positive $\exp(\omega^a)$ has the form $\omega^{\omega^{g(a)}}$ where $g(a) = \{c,g(a')\}|\{g(a'')\}$ and c is such that $a \sim \omega^c$. [It is under-

stood that 0 is not a possible value for a'.]

Let us apply theorem 10.11 to obtain $\exp(\omega^{\frac{1}{\omega}})$. First we evaluate $\exp \omega$. Since $\omega = \omega^1$ we have $a = 1$. Since $\omega^0 = 1$ we have $c = 0$. In this case we regard A' and A" as ϕ since $1 = \{0\}|\phi$ and 0 is separated out. Hence $g(1) = \{0\}|\phi = 1$. Thus $g(1) = 1$. Therefore $\exp \omega = \omega^{\omega^1} = \omega^\omega$ which agrees with the result we obtained by the earlier methods. We use induction to obtain $g(\frac{1}{2^n})$. Assume $g(\frac{1}{2^n}) = \frac{1}{2^n}$. Then

$g(\frac{1}{2^{n+1}}) = g[\{0\}|\{(\frac{1}{2^n})\}] = \{\text{ind } (\frac{1}{2^{n+1}})\}|\{g(\frac{1}{2^n})\} = \{0\}|\{\frac{1}{2^n}\} = \frac{1}{2^{n+1}}$. Therefore $g(\frac{1}{\omega}) = g[\{0\}|\{\frac{1}{2^n}\}] = \{\text{ind } \frac{1}{\omega}\}|\{g(\frac{1}{2^n})\} = \{-1\}|\{\frac{1}{2^n}\} = 0$. Hence $\exp(\omega^{\frac{1}{\omega}}) = \omega^{\omega^0} = \omega$ which again agrees with what we obtained earlier.

We show next a kind of converse; namely that $\ln(\omega^{\omega^a})$ is a power of ω for all a. In a way this is redundant since it will also follow from later results. However, the proof is of interest since it leads to an explicit inductive formula for the inverse of g. For this we need a lemma which states that the uniformity theorem is valid for the function $\ln x$. This result probably deserves to be a theorem but since we have no lemmas in this chapter so far and since we are thus already top heavy with theorems we leave this as a lemma. Besides, we are interested primarily in the function $\exp x$, so that the function $\ln x$ is introduced primarily to help us prove results about $\exp x$, e.g. $\exp x$ is onto the set of all positive surreals.

Lemma 10.1. The uniformity theorem is valid for $\ln x$.

Proof. Recall that this means that we can obtain $\ln(\omega^a)$ by using any representation of a in the form $a = F|G$ instead of being restricted to the canonical representation. As in all proofs of such theorems in which we use the inverse cofinality theorem followed by the cofinality theorem we need inequalities of a certain kind. In this case we have

$$\ln(\omega^b) = \{\ln(\omega^{b'})+n,\ \ln(\omega^{b''}) - \omega^{\frac{b''-b}{n}}\}|\{\ln(\omega^{b''})-n,\ \ln(\omega^{b'})+\omega^{\frac{b-b'}{n}}\}.$$

If $b' < x < b$ then $\ln(\omega^x)+n \geq \ln(\omega^{b'})+n$ and

$\ell n(\omega^x) + \omega^{\frac{b-x}{n}} \leq \ell n(\omega^{b'}) + \omega^{\frac{b-b'}{n}}$.

If $b < y < b''$ then $\ell n(\omega^y)+n \leq \ell n(\omega^{b''})+n$ and $\ell n(\omega^y) - \omega^{\frac{y-b}{n}} \geq \ell n(\omega^{b''}) - \omega^{\frac{b''-b}{n}}$.

The first inequalities of each pair are obvious since $\ell n(\omega^x)$ is an increasing function of x. The proofs of the second ones in each pair are similar. First $\omega^{\frac{b-b'}{n}} - \omega^{\frac{b-x}{n}} > \omega^{\frac{x-b'}{n}}$ since $\omega^{\frac{b-b'}{n}}$ has a higher order of magnitude than either $\omega^{\frac{b-x}{n}}$ or $\omega^{\frac{x-b'}{n}}$. Also $\omega^{\frac{x-b'}{n}} > \ell n(\omega^x) - \ell n(\omega^{b'})$. Hence $\ell n(\omega^x) + \omega^{\frac{b-x}{n}} < \ell n(\omega^{b'}) + \omega^{\frac{b-b'}{n}}$. Similarly $\omega^{\frac{b''-b}{n}} - \omega^{\frac{y-b}{n}} > \omega^{\frac{b''-y}{n}} > \ell n(\omega^{b''}) - \ell n(\omega^y)$. Hence $\ell n(\omega^y) - \omega^{\frac{y-b}{n}} > \ell n(\omega^{b''}) - \omega^{\frac{b''-b}{n}}$.

<u>Theorem 10.12</u>. For all a, $\ell n(\omega^{\omega^a})$ is a power of ω.

<u>Proof</u>. The lemma allows us to apply the formula in the definition to the representation $\{0,\omega^{a'}r\}|\{\omega^{a''}s\}$ of ω^a. Hence

$$\ell n(\omega^{\omega^a}) = \{\ell n(\omega^0)+n,\ \ell n(\omega^{\omega^{a'}}r)+n,\ \ell n(\omega^{\omega^{a''}}s) - \omega^{\frac{\omega^{a''}s-\omega^a}{n}}\}|$$

$$\{\ell n(\omega^{\omega^{a''}}s)-n,\ \ell n(\omega^0)+\omega^{\frac{\omega^a-0}{n}},\ \ell n(\omega^{\omega^{a'}}r) + \omega^{\frac{\omega^a-\omega^{a'}r}{n}}\}.$$

We now simplify. As a start we can get rid of $\ell n(\omega^0)$, which is 0, thus getting n as a lower term. Since we may assume that r is an integer we have $\ell n(\omega^{\omega^{a'}}r) = r\ell n(\omega^{\omega^{a'}})$. Note that the addition formula for the function $\ell n\ x$ follows from the corresponding property for $\exp x$ since we already know that it is the inverse function. Note that if $x > 0$ $\ell n(\omega^x)$ is positive infinite. Hence $\ell n(\omega^{\omega^{a'}})$ is infinite so $r\ell n(\omega^{\omega^{a'}})$ is mutually cofinal with $r\ell n(\omega^{\omega^{a'}})+n$.

Now consider $\frac{\omega^{a''}s-\omega^{a}}{n}$ which occurs as an exponent. Clearly $\frac{\omega^{a''}s}{n+1} < \frac{\omega^{a''}s-\omega^{a}}{n}$. Since $\ell n(\omega^{\omega^{a''}}s) < \omega^{\frac{\omega^{a''}s}{n}}$ this shows that all the lower terms in the third group are negative and hence can be discarded because of the presence of n. So the set of lower terms have been simplified to $\{n, r\ell n(\omega^{\omega^{a'}})\}$. Similarly for the upper terms we may assume that s is dyadic and drop n so that the first group of terms simplify to $s\ell n(\omega^{\omega^{a'}})$. The second group of terms have the form $\omega^{\frac{\omega^{a}}{n}}$. Now consider $\frac{\omega^{a}-\omega^{a'}r}{n}$ which occurs as an exponent. $\frac{\omega^{a}\ \omega^{a'}r}{n} > \frac{\omega^{a}}{n+1}$. Since $\ell n(\omega^{\omega^{a'}}r) > 0$ we may discard the third troup of terms by cofinality because of the second group. Thus the set of upper terms simplify to $\{s\ell n(\omega^{\omega^{a''}}), \omega^{\frac{\omega^{a}}{n}}\}$. Again, since $\omega^{\frac{\omega^{a}}{n+1}} < s\omega^{\frac{\omega^{a}}{n}}$ we may again by cofinality slightly unsimplify the set of upper terms to $\{s\ell n(\omega^{\omega^{a''}}), s\omega^{\frac{\omega^{a}}{n}}\}$.

The final representation now exhibits $\ell n(\omega^{\omega^{a}})$ as a surreal of the form ω^{x}. In fact, if we define $h(x)$ so that $\ell n(\omega^{\omega^{x}}) = \omega^{h(x)}$ then $h(a) = \{0, h(a')\}|\{h(a''), \frac{\omega^{a}}{n}\}$.

Note that $h(a) > 0$ for all a. This shows that the range of g consists of the class of <u>all</u> surreal numbers. Thus $\exp x$ induces a map from the class of positive surreals <u>onto</u> the class of all surreals! This sounds like the opposite of what an exponential function does but recall that the map goes from exponents to iterated exponents so that there is nothing unreasonable about that, although the existence of such a g is a surprising phenomenon.

The uniformity theorem is valid for g and h. This is clear since $x \leq y \to \text{ind } x \leq \text{ind } y$. We are primarily interested in g and it is thus convenient to know that we may use any representation of b to compute $g(b)$ from the formula given by theorem 10.11.

We next prove that $\exp x$ has a kind of generalized linearity property. This will enable us to reduce the study of $\exp x$ for purely infinite x entirely to the study of the function g.

Theorem 10.13. If $a_i > 0$ for all i then
$\exp \sum_{i<\alpha} \omega^{a_i} r_i = \omega^y$ where $y = \sum_{i<\alpha} \omega^{g(a_i)} r_i$.

Proof. Since $\exp(x+y) = \exp x \exp y$ and $\omega^{x+y} = \omega^x \omega^y$ the result is clear for finite sums and rational r_i. We shall prove first that the result is valid for monomials for arbitrary real r_i and then prove the result for arbitrary sums.

Consider $\omega^a r$ where r is real. Then $\omega^a r$ may be expressed as $\{\omega^a r'\}|\{\omega^a r''\}$ where all elements of the form r' and r'' are dyadic, in particular rational, so that the result is known for $\omega^a r'$ and $\omega^a r''$. Then $\exp(\omega^a r) = \{0, \exp(\omega^a r')(\omega^a r - \omega^a r')n\}|\{\frac{\exp(\omega^a r'')}{(\omega^a r'' - \omega^a r)^n}\}$.
Since the set of elements of the form r' is non-empty we can eliminate 0. Since $(\omega^a r - \omega^a r')^{n+2} > \omega^{a(n+1)} > (\omega^a r - \omega^a r')^n$ we may simplify the set of lower elements by mutual cofinality and similarly for the upper elements so that we obtain

$$\exp(\omega^a r) = \{0, \omega^{\omega^{g(a)} r'} \omega^{na}\}|\{\omega^{\omega^{g(a)} r''} \omega^{-na}\}$$
$$= \{0, \omega^{\omega^{g(a)} r' + na}\}|\{\omega^{\omega^{g(a)} r'' - na}\}.$$

Now $\exp(\omega^a) > \omega^{na}$, i.e. $\omega^{\omega^{g(a)}} > \omega^{na}$ thus $\omega^{g(a)} > na$ in general. Since this is true for all positive integers we also have $\omega^{g(a)} > (\frac{n}{r-r'})a$, i.e. $\omega^{g(a)} r - \omega^{g(a)} r' > na$ hence $\omega^{g(a)} r' + na < \omega^{g(a)} r$. Similarly $\omega^{g(a)} r'' - na > \omega^{g(a)} r$. This proves that $\omega^{\omega^{g(a)} r}$ satisfies the betweenness condition with respect to the representation of $\exp(\omega^a r)$. Now $\omega^{g(a)} r$ can be expressed as $\{\omega^{g(a)} r'\}|\{\omega^{g(a)} r''\}$. Since the lower terms have no maximum and the uppper terms no minimum we have

$$\omega^{\omega^{g(a)} r} = \{0, \omega^{\omega^{g(a)} r'}\} \{\omega^{\omega^{g(a)} r''}\}.$$

Certainly $\omega^{g(a)} r' + na \geq \omega^{g(a')} r'$ and $\omega^{g(a)} r'' - na \leq \omega^{g(a)} r''$ so cofinality is immediate, completing the proof in the case for monomials.

For arbitrary sums we use induction on α. For non-limit ordinals the result follows immediately from the additive properties of the functions $\exp x$ and ω^x since the problem reduces to one of ordinary addition. For limit ordinals we need an argument which is roughly similar to the one used before for monomials.

$$\sum_{i<\alpha} \omega^{a_i} r_i = \{\sum_{i\leq\gamma} \omega^{a_i} r_i - \omega^{a_\gamma}\varepsilon\}|\{\sum_{i\leq\gamma} \omega^{a_i} r_i + \omega^{a_\gamma}\varepsilon\}$$

where $\gamma < \alpha$ and ε is positive. Hence

$$\exp(\sum_{i<\alpha} \omega^{a_i} r_i) = \{0, \exp(\sum_{i\leq\gamma} \omega^{a_i} r_i - \omega^{a_\gamma}\varepsilon)(\omega^{a_\gamma}\varepsilon')^n\}|$$

$\{\exp(\sum_{i\leq\gamma} \omega^{a_i} r_i + \omega^{a_\gamma})(\omega^{a_\gamma}\varepsilon'')^{-n}\}$ where ε' is such that

$\omega^{a_\gamma}\varepsilon' = \omega^{a_\gamma}\varepsilon + \sum_{\gamma<i<\alpha} \omega^{a_i} r_i$ and similarly for ε''. $|\varepsilon-\varepsilon'|$ and $|\varepsilon-\varepsilon''|$ are infinitesimal. Also

$$\sum_{i<\alpha} \omega^{g(a_i)} r_i = \{\sum_{i\leq\gamma} \omega^{g(a_i)} r_i - \omega^{g(a_\gamma)}\varepsilon\}|\{\sum_{i\leq\gamma} \omega^{g(a_i)} r_i + \omega^{g(a_\gamma)}\varepsilon\}.$$

Since the lower terms have no maximum and the upper terms no minimum we have $\omega^{\sum_{i<a} \omega^{g(a_i)} r_i} = \{0,\omega^F\}|\{\omega^G\}$ where F stands for the set of lower terms and G for the set of upper terms.

We now use cofinality to show that the representation of $\exp(\sum_{i<\alpha} \omega^{a_i} r_i)$ does give $\omega^{\sum_{i<a} \omega^{g(a_i)} r_i}$.

First we verify the betweenness condition. By mutual cofinality terms such as $(\omega^{a_\gamma}\varepsilon')^n$ may be replaced by ω^{na_γ}. Now we know that $\omega^{g(a)} > na$ for all integers n. Since ε is not infinitesimal we have $\omega^{g(a)}\frac{\varepsilon}{2} > na$ for all integers n. Now a typical term among the lower terms of $\exp(\sum_{i<\alpha} \omega^{a_i} r_i)$ is $\exp(\sum_{i\leq\gamma} \omega^{a_i} r_i - \omega^{a_\gamma}\varepsilon)\omega^{na_\gamma}$. This is of the form ω^y where $y = \sum_{i\leq\gamma} \omega^{g(a_i)} r_i - \omega^{g(a_\gamma)}\varepsilon + na_\gamma$ using the inductive hypothesis and addition theorem for the function $\exp x$. This in turn is smaller then

$\sum_{i<\gamma} \omega^{g(a_i)} r_i - \omega^{g(a_\gamma)}\varepsilon + \omega^{g(a_\gamma)\varepsilon}\frac{}{2}$ which is less than $\sum_{i<\alpha} \omega^{g(a_i)} r_i$ by the lexicographical order. This is exactly what we need. A similar argument applies to the upper terms.

Cofinality is immediate since a typical term of ω^F has the form $\exp(\sum_{i<\gamma} \omega^{a_i} r_i - \omega^{a_\gamma}\varepsilon)$ by the inductive hypothesis, thus all we need is the fact that $\omega^{na} \geq 1$. The situation is similar for terms of the form ω^G. This completes the proof.

The study of the function $\exp x$ has now been reduced to the study of g. Recall from theorem 10.11 that $g(a) = \{ind(a), g(a')\}|\{g(a'')\}$ where $ind\ a$ is the unique c such that $a \sim \omega^c$ Thus, as has been promised earlier, the subject has in one respect become greatly simplified. On the other hand, g behaves in tricky interesting ways so that the subject is still far from trivial.

D PROPERTIES OF g AND EXPLICIT RESULTS

We shall first determine g for ordinals. We know already that $g(1) = 1$. It is easy to see that $g(2) = 2$, $g(3) = 3, \ldots, g(\omega) = \omega$, $g(\omega^2) = \omega^2$, $g(\omega^\omega) = \omega^\omega$, etc. We thus have an excellent physicist's proof that $g(x) = x$ for all ordinals which is similar but somewhat superior to the physicist's proof that all odd numbers are prime! At least in this case the first counter-example is not a finite number such as 9.

Before keeping the reader in too much suspense let us consider an epsilon number ε. By definition $\omega^\varepsilon = \varepsilon$ hence $ind\ \varepsilon = \varepsilon$. However, $ind\ \varepsilon$ appears as a lower element in the formula for $g(\varepsilon)$. Hence $g(\varepsilon) > \varepsilon$. This can also be seen from the formula for $\exp x$. First, we have $\omega^\varepsilon = \varepsilon$ and $\omega^{\omega^\varepsilon} = \varepsilon$. Now $\exp(x) > x^n$ for positive purely infinite numbers. So certainly $\exp \varepsilon > \varepsilon$, i.e. $\exp(\omega^\varepsilon) > \omega^{\omega^\varepsilon}$.

Thus the subject of the exponential function has taken on a new twist. We shall see that epsilon numbers and numbers related to them play an important part in the determination of an explicit formula for g.

Incidentally, it is impossible to give an explicit definition of "explicit." The reader may claim that theorem 10.11 already gives a definition. However, we shall see that we can obtain results which are much more explicit, although unlike the earlier part of the theory in this chapter there is a feeling of lack of finality in some of the results. For example, what we do not do in general is obtain the sign sequence of $g(x)$ directly from the sign sequence of x though in a sense we approximate this in some respects to an extent which is conveniently tractable and interesting.

First, we obtain a formula for $g(a)$ in the special case where a is an ordinal.

Theorem 10.14. If a is an ordinal, then $g(a) = a$ unless a satisfies the inequality $\varepsilon \leq a < \varepsilon + \omega$ for some epsilon number ε, in which case $g(a) = a+1$.

Proof. If a is an ordinal $g(a)$ has the form $\{\text{ind } a, g(a')\}|\phi$ since a'' is empty. Hence $g(a)$ is an ordinal. We now use induction.

First suppose that a is less than the first epsilon number. Then $\text{ind } a < a$. By the inductive hypothesis $g(a) = \{\text{ind } a, a'\}|\phi$. Since $\text{ind } a < a$ this is a by the cofinality theorem.

Now let a be the first epsilon number. Then $g(a) = \{\text{ind } a, a'\}|\phi = \{a,a'\}|\phi = a+1$.

We now determine $g(a+n)$ inductively on n where n is a positive integer $g(a+n+1) = \{\text{ind}(a+n+1), g(a+n)\}|\phi$ by cofinality. Now $\text{ind}(a+n+1) = a$. Hence we obtain $\{a,a+n+1\}|\phi = a+n+2$.

Next we determine $g(a+\omega)$. In fact, $g(a+\omega) = \{\text{ind}(a+\omega), g(a+n)\}|\phi = \{a, a+n+1\}|\phi = a+\omega$. Thus we get back to the equality $g(x) = x$. It is now easy to see that $g(x) = x$ from now on until we get to the next epsilon number. The general argument is similar to what we have so far. In fact, suppose the theorem is valid for all a such that $a \leq \varepsilon_\alpha$ where ε_α is the αth epsilon number. Then by the same argument as for the first epsilon number we have by induction on n that $g(\varepsilon_\alpha+n+1) = \{\text{ind}(\varepsilon_\alpha+n+1), g(\varepsilon_\alpha+n\}|\phi = \{\varepsilon_\alpha,\varepsilon_\alpha+n+1\}|\phi = \varepsilon_\alpha+n+2$ and $g(\varepsilon_\alpha+\omega) = \{\varepsilon_\alpha,\varepsilon_\alpha+n+1\}|\phi = \varepsilon_\alpha+\omega$. If $x > \omega$ but $\varepsilon_\alpha+x < \varepsilon_{\alpha+1}$ then we have $g(\varepsilon_\alpha+x) = \{\text{ind}(\varepsilon_\alpha+x),(\varepsilon_\alpha+x)'\}|\phi$. Since $\text{ind}(\varepsilon_\alpha+x) < \varepsilon_\alpha+x$ this is $\varepsilon_\alpha+x$ by cofinality. Then $g(\varepsilon_{\alpha+1}) = \{\text{ind}(\varepsilon_{\alpha+1}),(\varepsilon_{\alpha+1})'\}| = \varepsilon_{\alpha+1}+1$.

To complete the proof we still must consider ε_α where α is a limit ordinal, but this is also more or less similar to what we had previously. In fact, by cofinality $\varepsilon_\alpha = \{\varepsilon_\alpha'\}|\phi$, hence

$g(\varepsilon_\alpha) = \{\text{ind } \varepsilon_\alpha, g(\varepsilon_\alpha')\}|\phi = \{\varepsilon_\alpha, \varepsilon_\alpha'+1\}|\phi = \varepsilon_\alpha+1$ since certainly $\varepsilon_\alpha'+1 < \varepsilon_\alpha$ for any α'.

The above theorem is the first example of a phenomenon which occurs in the study of g. Essentially we have a kind of "singularity" in the neighborhood of an epsilon number. This helps to make the study of g somewhat tricky.

We now consider the special case where a consists of a plus followed by a sequence of minuses. These sequences lead to arbitrarily small positive surreals and is thus a natural class of surreals to consider after the ordinals, which lead to arbitrarily large surreals.

Actually we already computed the result when the number of minuses is finite or exactly ω. In fact, we have $g(\frac{1}{2^n}) = \frac{1}{2^n}$ and $g(\frac{1}{\omega}) = 0$. We note here that the pattern $g(x) = x$ is broken at $\frac{1}{\omega}$ in a different way than at epsilon numbers.

Now by the sign sequence formula $2^{-n}\omega^{-b}$ where n is a positive integer and b is an ordinal consists of a plus followed by $\omega b+n$ minuses. Thus we are interested in a formula for $g(2^{-n}\omega^{-b})$.

In the sequel we expect the reader to have some facility in computing expressions $F|G$ from the sign sequences of elements in F and G. This is an essential skill for the study of the function g.

Theorem 10.15. $g(2^{-n}\omega^{-b}) = -b + 2^{-n}$.

Proof. We use induction on b and induction on n for fixed b. Suppose $g(2^{-n}\omega^{-b}) = -b + 2^{-n}$. Consider $2^{-n-1}\omega^{-b}$. This is $\{0\}|\{2^{-n}\omega^{-b}\}$ by the sign sequence formula and cofinality. (This can also be seen by other methods from chapter five, but it is easiest to quote the sign sequence formula.) Hence

$g(2^{-n-1}\omega^{-b}) = \{\text{ind } 2^{-n-1}\omega^{-b}\}|\{g(2^{-n}\omega^{-b})\} = \{-b\}|\{-b+2^{-n}\} = -b+2^{-n-1}$.

$g(\omega^{-b}) = \{\text{ind}(\omega^{-b}\}|\{g(2^{-n}\omega^{-b'})\} = \{-b\}|\{-b'+2^{-n}\} = -b+1 = -b+2^{-0}$. So the formula is valid in this case.

Remark. Note that $-b+1$ is simply the negative of an ordinal when b is a non-limit ordinal but is different when b is a limit ordinal. In fact, if b is a non-limit ordinal $-b+1$ consists of $b-1$ minuses, whereas if b is a limit ordinal $-b+1$ consists of b minuses followed by a plus. In either case the computation $\{-b\}|\{-b'+2^{-n}\} = -b+1$ is valid. (Note that $-b$ consists of a sequence of b minuses and $-b'+2^{-n}$ consists of a sequence of less than b minuses followed by a plus and a finite number of minuses.)

The next case of interest is an ordinary ε number followed by a sequence of minuses. This case has some resemblance to the case just considered. Here we feel that it is more instructive to use a somewhat more informal approach.

Now clearly ind b is a monotonic function of b. As we add on more and more minuses to ε, ind b remains constant until we reach a certain point at which the value is decreased. We first investigate the pattern while ind b is constant and then determine when ind b changes and the effect of the change on $g(b)$.

We know that $g(\varepsilon) = \varepsilon+1$ which is ε followed by a plus. If b consists of ε followed by a sequence of α minuses then the lower elements of $g(b)$ have the form $\{\text{ind } b, x\}$. Since $\text{ind } b = \varepsilon$ and $x < \varepsilon$ we may simply replace this by ε. The upper elements of $g(b)$ have the form $g(x)$ where x consists of ε followed by β minuses where $\beta < \alpha$. It now follows trivially by induction that $g(b)$ consists of ε followed by a plus and α minuses. This is an example of the convenient fact which we shall see in a more general form later that as long as ind b does not change, then roughly speaking $g(b)$ continues the same way as b. Complications are caused by changes in ind b although not all changes in ind b cause problems. In fact, we have already seen that for ordinals tricky changes occur at epsilon numbers although indices change at many other ordinals. (For example, although ind b changes at ω^2, this causes no complication in the computation of $g(\omega^2)$.)

We now study the variation in ind b as the number of minuses is increased. For this we need the sign sequence formula. In fact, if c is ε followed by α minuses, then ω^c is ε followed by $\varepsilon\omega\alpha$ minuses and more generally $2^{-n}\omega^c$ is ε followed by $\varepsilon\omega c + \varepsilon n$ minuses. This is enough information for the determination of ind b. In fact, if b is ε followed by β minuses, then ind b is ε followed by α minuses where α is the quotient obtained by dividing β by $\varepsilon\omega$ using the division algorithm; i.e. informally speaking, as minuses are added to b, minuses are added to ind b at intervals of length $\varepsilon\omega$.

So the above description of $g(b)$ applies if the number of minuses after ε is less than $\varepsilon\omega$. If there are exactly $\varepsilon\omega$ minuses after ε, then ind b is ε followed by a minus. The set of lower elements now have the form $\{\varepsilon-1,x\}$. The upper elements all consist of ε followed by a plus and a sequence of minuses. It is clear by cofinality that $g(b) = \varepsilon$, i.e. $g(\omega^{\varepsilon-1}) = \varepsilon$. We now have a setup for a double induction.

We lead the reader by the hand for a short distance. If b has $\varepsilon\omega+1$ minuses after ε then we may express $g(b)$ as $\{\varepsilon-1\}|\{\varepsilon\}$ by cofinality. Hence $g(b)$ consists of ε followed by a minus and a plus. $\varepsilon-1$ remains as a lower element until we obtain $\varepsilon\omega\cdot 2$ minuses; hence as we noted earlier $g(b)$ continues the same way as b. Hence if b has $\varepsilon\omega+r$ minuses after ε where r is finite, then $g(b)$ is ε followed by a minus and plus followed by $r-1$ minuses. If r is infinite then $g(b)$ also has a tail of r minuses since the shortage by one gets wiped out at $\varepsilon\omega + \omega$.

When b has $\varepsilon\omega\cdot 2$ minuses after ε, $\varepsilon-1$ no longer occurs as a lower element and we thus easily obtain $g(b) = \varepsilon-1$. The situation for $\varepsilon\omega\cdot 2 + r$ is similar to what we had before, the only difference being that $g(b)$ has two minuses after ε rather than just one. This continues in a similar way up to any b which consists of ε followed by $\varepsilon\omega n + r$ minuses. $g(b)$ consists of n minuses after ε followed by a plus and the contribution of r.

We now write $\beta = \varepsilon\omega\alpha + r$. The induction on α works the same way for all non-limit ordinals. In fact, if $\beta = \varepsilon\omega\alpha$ then $g(b)$ consists of ε followed by $\alpha-1$ minuses. If $\beta = \varepsilon\omega\alpha + r$ for finite r then $g(b)$ consists of ε followed by α minuses, a plus, and $r-1$ minuses. If $\beta = \varepsilon\omega\alpha + r$ for infinite r then $g(b)$ consists of ε followed by α minuses, a plus, and r minuses. A slight difference occurs if α is a limit ordinal. The case $\beta = \varepsilon\omega^2$ is typical. In this case ind b which is ε followed by ω minuses is cofinal in the set of lower elements. For the upper elements we have the cofinal set consisting of the elements of the form ε followed by n minuses. Hence $g(b)$ consists of ε followed by ω minuses and a plus. If $\beta = \varepsilon\omega^2 + r$ then $g(b)$ consists of ε followed by ω minuses, a plus, and r minuses. Note that for finite r this is different from the case where α is a non-limit ordinal because of the start at $r = 0$. In general, if $\beta = \varepsilon\omega\alpha + r$ where α is a limit ordinal $g(b)$ consists of ε followed by α minuses, a plus, and r minuses.

Note that if b is ε followed by $\varepsilon\omega\alpha + \varepsilon m + r$ minuses then b has the form $2^{-m}\omega^{y} - r$ where y is ε followed by α minuses. The r minuses in the tail of b contribute $-r$ to the value of b. On the other hand, the corresponding minuses in $g(b)$ contribute something entirely different. For example, if $g(b)$ consists of ε followed by a plus and r minuses then $g(b) = \varepsilon + y$ where y is the sequence consisting of a plus followed by r minuses. We already noted just before the proof of theorem 10.15 that y is of the form $2^{-n}\omega^{-a}$.

The above computation does not appear to be too interesting in general. It seems that the most succinct way to look at the behavior of $g(b)$ is to describe b in terms of a normal form followed by a sequence of pluses and minuses. We shall see that some nice general results exist if b is expressed in this form. On the other hand, attempts to deal with b described purely in algebraic form or purely as a sign sequence lead rapidly to messy computations.

We now turn to a study of more general b. We shall see that the special cases considered are basic building blocks for such a study.

First, we have a rather general inequality.

Theorem 10.16. If $b \geq 1$ then $g(b) \geq b$.

Proof. By theorem 10.14 this is true if b is an ordinal. (In fact, the theorem gives more precise information concerning $g(b)$.) We use induction on the length of b. Since we know the result for $b = 1$ we may assume that $b > 1$. Hence b begins with at least two pluses. Therefore in the canonical representation of b all lower terms b' begin with at least two pluses unless $b' = 0$ or 1. Since 0 is discarded in the computation of $g(b)$, we may thus assume that $b' \geq 1$, hence $g(b') \geq b'$. Certainly every upper term b'' satisfies $b'' \geq 1$, hence $g(b'') \geq b''$.

Now $g(b) = \{\text{ind } b, g(b')\}|\{g(b'')\}$. Of course we have $b = \{b'\}|\{b''\}$. We now apply theorem 2.5. $b < b'' \leq g(b'')$. Also $b' \leq g(b') < g(b)$ if $b' \neq 0$. If $b' = 0$ we have $0 \leq g(1) < g(b)$, hence $b \leq g(b)$.

Remarks. Although the cofinality theorems have been frequently used throughout the book this is our first and only use of theorem 2.5. It happens to be convenient to use this theorem here since we are proving an inequality.

There is a technicality to watch for. If $b' = 0$ then there is no $g(b')$. In our proof we used $g(1)$, which is legal, since 1 is a lower element of b. If one is not cautious about this, one can get a false proof that $g(b) \geq b$ for all b.

The requirement $b \geq 1$ can be slightly weakened. We shall sharpen theorem 10.16 later. At any rate, we know that $g(\frac{1}{\omega}) = 0$ so $g(b) < b$ if $b \leq \frac{1}{\omega}$. This gives us a limitation as to how much the hypothesis can be weakened. Actually, we may replace $b \geq 1$ by $b \geq 2^{-n}$ for all positive integers n thanks to the fact that $g(2^{-n}) = 2^{-n}$. On the other hand, this does not give the ultimate in sharpness since there is still a question concerning infinitesimals which are larger than $\frac{1}{\omega}$.

We now generalize part of theorem 10.14 from ordinals to more general b.

Theorem 10.17. If $\varepsilon_\alpha+\omega \leq b \leq \alpha < \varepsilon_{\alpha+1}$ for some epsilon number ε_α and some ordinal α then $g(b) = b$. The same conclusion applies if $1 \leq b \leq \alpha < \varepsilon_0$.

Note. The inequality $b \leq \alpha < \varepsilon_{\alpha+1}$ cannot be replaced by $b < \varepsilon_{\alpha+1}$. In fact, if b has the form $\varepsilon-1$ then we already know that $g(b) > b$.

Proof. Since we already have the result when b is an ordinal we may assume that b is not an ordinal, so $\varepsilon_\alpha+\omega < b < \alpha$. We use induction. Let β be the least ordinal larger than b. Then $\varepsilon_\alpha+\omega < b < \beta$. Since b is not an ordinal, it follows from the definition of β that b consists of β followed by a minus and possibly other terms. So β is an upper element in the canonical representation of b and all other upper elements are less than β. For the lower elements we may assume by cofinality that they are all at least $\varepsilon_\alpha+\omega$. Hence all numbers in the representation of b satisfy the hypothesis of the theorem so that by the inductive hypothesis $g(b) = \{\text{ind } b, g(b')\}|\{g(b'')\} = \{\text{ind } b, b'\}|\{b''\}$. (As a technical point, we may reinsert all ordinals which are less than

$\varepsilon_\alpha+\omega$ into the set of all b' so that the lower set $\{\text{ind } b, b'\}$ has its usual meaning.)

Now β is not an epsilon number since it is strictly between ε_α and $\varepsilon_{\alpha+1}$. Hence ind β is an ordinal less than β. Therefore ind $b \leq$ ind $\beta < b$. Therefore $\{\text{ind } b, b'\}|\{b''\} = b$ by cofinality. This completes the proof.

The same argument applies to the second part. As we noted earlier, as a technical point we need the element one as a lower element so that we can reintroduce zero by cofinality. Formally speaking, if one is picayune and use the notation B' for the set of all b', the set of lower elements should be expressed first as $\{\text{ind } b\},(B'-\{0\})$ which simplifies to $\{\text{ind } b\},\{B'\}$ by cofinality. We prefer a slight abuse of notation for the sake of readability.

The above theorem shows that the equality $g(b) = b$ is par for the course, i.e in a rough sense "most" numbers are covered by the theorem. We now consider numbers which are "close" to epsilon numbers.

Theorem 10.18. If $\varepsilon \leq b \leq \varepsilon+n$ for some epsilon number ε and integer n then $g(b)$ is obtained from b as follows. If the sign sequence of b is ε followed by the sequence S then $g(b)$ is the juxtaposition ε followed by a plus and then S.

Note. By a careful look at the sign sequence formula one can show that $g(b) = b+1$, as was stated as part of theorem 10.14 in the special case where b is an ordinal. We prefer to deemphasize this point of view since in general a simple operation such as inserting a plus into a sequence can have a complicated effect on the normal form. This is because the existence of the plus affects the contribution of the subsequent signs. Although the general problem concerning the effect of insertion of signs on the algebraic value of a surreal number is of intrinsic interest, it is far removed from the theory of exponentiation. For our purpose it suffices to attempt to express the function g in what appears to be the most tractable way.

Proof. The result is known for ordinals. We may thus assume that for some integer $n>0$ we have $b = \varepsilon_\alpha+n+c$ where $-1<c<0$, i.e. b begins with $\varepsilon+n$ pluses followed by a minus. Hence a cofinal lower set in the canonical representation of b may be obtained by restricting to numbers which begin with $\varepsilon+(n-1)$ pluses. All upper elements begin with $\varepsilon+n$ pluses and except for $\varepsilon+n$ itself continue with a minus. Thus b' and b'' satisfy the hypothesis of the theorem so we can use induction. Since $\text{ind}\ \varepsilon = \text{ind}(\varepsilon+w) = \varepsilon$ then $\text{ind}\ b = \varepsilon$. Also, since $g[\varepsilon+n-1] = \varepsilon+n$ is a lower element of $g(b)$, $\text{ind}\ b = \varepsilon$ is superfluous by cofinality. Hence $g(b) = \{g(b')\}\{g(b'')\}$. By the inductive hypothesis all the elements of the form $g(b')$ and $g(b'')$ have a plus inserted after ε. By the lexicographical order $g(b)$ also has this property.

We now fill one of the gaps in our study of b and strengthen theorem 10.17 slightly.

Theorem 10.19. If for some epsilon number ε and all integers n, $\varepsilon+n \leq b$ but $b < \varepsilon+\omega$, then $g(b) = b$.

Note. This takes care of one of the gaps left by theorems 10.17 and 10.18.

Proof. Such a b begins with $\varepsilon+\omega$ followed by a minus. $g(b) = \{\text{ind}\ b\ g(b')\}|\{g(b'')\}$. $\text{ind}\ b = \varepsilon$ and since $g(\varepsilon+n) = \varepsilon+n+1$ is a lower element, ε is superfluous. Hence $g(b) = \{g(b')\}|\{g(b'')\}$.

Assume first that after the $\varepsilon+\omega$ pluses and minus there are no further changes in sign. We then set up an easy induction. If there is only one minus after the $\varepsilon+\omega$ pluses then $g(b) = \{g(b')\}|\{g(b'')\} = \{\varepsilon+n+1\}|\{\varepsilon+\omega\} = \{\varepsilon+n\}|\{\varepsilon+\omega\} = b$. An easy induction gives the result in this case.

If there are further changes in sign then we have lower elements b' which satisfy $g(b') = b'$ by the previous case. So using induction we have a cofinal set of lower elements satisfying $g(b') = b'$, so $g(b) = \{g(b')\}|\{g(b'')\} = \{b'\}|\{b''\} = b$.

There is only one "border" area left. Namely, this is the one where b begins with ε pluses for an epsilon number ε followed by a minus, i.e. $b < \varepsilon$ but for no ordinal $\beta < \varepsilon$ does b satisfy $b \leq \beta$. This turns out to be the most complicated case. We include numbers less than 1 into this case, i.e. numbers beginning with a plus followed by a minus. Note that all other numbers have been handled by earlier theorems. These two classes of numbers behave in a similar manner. It seems strange to regard 1 as an epsilon number, yet when it comes to the study of g, 1 behaves enough like an epsilon number so that it is convenient for the sake of efficiency to study the classes simultaneously. So in future when we refer to an epsilon number followed by minuses it is understood that this includes the case where we have one plus followed by minuses.

In theorem 10.15 and in the discussion afterwards we already saw what happens when the continuation consists solely of minuses. The main characteristic of this case is that ind b changes as the number of minuses is increased at various intervals so that $g(b)$ changes in an abrupt way rather than continuing in the same manner as b.

Suppose that the number of minuses after ε before the next plus is expressed in the form $\varepsilon\omega\alpha+r$ where $r < \varepsilon\omega$. We now break this up into several subcases, only one of which is especially tricky.

Case 1. $r > \omega$. Here we obtain an upper element d in the canonical representation of b by taking ε followed by $\varepsilon\omega\alpha+\omega$ minus and a lower element c by taking ε followed by $\varepsilon\omega\alpha+r$ minuses. Now ind x is constant for $c \leq x \leq d$. Furthermore $g(d)$ has the form p followed by ω minuses and $g(c)$ has the form p followed by r minuses. (In this case p consists of ε followed by α minuses followed by a plus. However, since the subsequent reasoning is independent of the specific nature of p it is convenient for the sake of generality to ignore this fact now.) Any time we have such a situation we can easily determine $g(x)$ by induction for any x satisfying $c \leq x \leq d$ by a type of argument we have already used several times. First, since ind x = ind $c < g(c)$ and c is a lower element in the canonical representation of x, it is superfluous to have ind x as a lower element by cofinality. By a further use of cofinality we may limit the upper and lower elements used in the representation to numbers y such that $c \leq y \leq d$. Then by an obvious induction and use of the lexicographical order we see that if any number x satisfying $c \leq x \leq d$ is expressed as the juxtaposition consisting of ε followed by $\varepsilon\omega\alpha+\omega$ minuses followed by q then $g(x)$ is the juxtaposition p followed by ω minuses followed by q.

Case 2. $r > 0$ and α is a limit ordinal. Then the same reasoning applies exactly providing ω is replaced by the empty sequence.

Case 3. $r = \omega$ and α is a non-limit ordinal. In this case we replace ω by any finite n larger than 0. We note that if d is the juxtaposition of ε followed by $\varepsilon\omega\alpha+n$ minuses then the last string of minuses in $g(d)$ contains only $n-1$ minuses. However, since n runs through the set of all positive integers in obtaining upper sums, this makes no essential difference.

Case 4. $r > 1$ and α is a non-limit ordinal. We can now replace ω by 1. This is otherwise like case 1 except that the r minuses in x give rise to $r-1$ minuses in $g(x)$.

Case 5. $r = 1$ and α is a non-limit ordinal. Here we let $d = \varepsilon$ followed by $\varepsilon\omega\alpha$ minuses and $c = \varepsilon$ followed by $\varepsilon\omega\alpha+1$ minuses. This is essentially the same as the other cases, the only difference being that the last minus in c contributes a minus followed by a plus in $g(c)$. For $c \leq x \leq d$ the sign sequence in $g(x)$ continues as in x. Note only that if there is a finite string of n pluses after the $\varepsilon\omega\alpha+1$ minuses in x, $g(x)$ will have a string of $n+1$ pluses because of the extra plus at the beginning.

At any rate all these cases in which $r > 0$ are essentially the same, differing only in very minor ways. In all cases the sign sequence of $g(x)$ continues as in x. The cases where $r = 0$ are more challenging. Roughly speaking, we are then closer to the "border" where ind x changes.

Case 6. $r = 0$ and α is a non-limit ordinal. This case also turns out not to be too complicated. Recall that if c is ε followed by $\varepsilon\omega\alpha$ minuses then $g(c)$ is ε followed by $\alpha-1$ minuses; and if d_r is ε followed by $\varepsilon\omega(\alpha-1)+r$ minuses where r is less than $\varepsilon\omega$ and non-zero, then $g(d_r)$ is ε followed by $\alpha-1$ minuses followed by a plus and $r-1$ or r minuses depending on whether r is finite or not. Here we have $c \leq x \leq d_r$ for all $r < \varepsilon\omega$. Ind x is not fixed in this interval. However, ind $x \leq$ ind $d_r = g(c)$, so if $x > c$ then ind x is superfluous in the representation of $g(x)$.

Now let $b = c$ followed by a plus. Then it is easy to see that $g(b) = \{g(c)\}|\{g(d_r\}$ which is ε followed by $\alpha-1$ minuses followed by a plus and $\varepsilon\omega$ minuses. In other words, the final plus contributes a plus followed by $\varepsilon\omega$ minuses. It is now easy to see that for any x satisfying $c \leq x \leq d_r$ the sign sequence of $g(x)$ for x beyond ε, the $\varepsilon\omega\alpha$ minuses, and plus continues as in x. So this case is much like the cases where $r > 0$, the main difference being the existence of a plus in x which contributes a plus followed by $\varepsilon\omega$ minuses in $g(x)$.

Case 7. $r = 0$ and α is a limit ordinal. This is the only case left to consider and the only case which is really fundamentally different. Let $c = \varepsilon$ followed by $\varepsilon\omega\alpha$ minuses and $d_\beta = \varepsilon$ followed by $\varepsilon\omega\beta$ minuses where $\beta < \alpha$. $g(c)$ is ε followed by α minuses and a plus. We may assume by cofinality that β is a non-limit ordinal in which case $g(d_\beta)$ is ε followed by $\beta-1$ minuses. We are now interested in the interval $c \leq x \leq d_\beta$ for all $\beta < \alpha$.

As in the last case let $b = c$ followed by a plus. Then $g(b) = \{\text{ind } b, g(c)\}|\{g(d_\beta)\}$. It is immediate that this is ε followed by α minuses and two pluses. (Note that we use the fact that ind b = ind c since an extra plus cannot change the value of ind.)

So far this case looks easier than the previous one since the extra plus contributes simply a plus. One can almost say that for an arbitrary x satisfying $c \leq x \leq d_\beta$ the sign sequence for the rest of $g(x)$ continues as in x. However, we must beware of the contribution of ind x.

In fact, suppose that after ε and the $\varepsilon\omega\alpha$ minuses we have only pluses. The variation in ind x as the number of pluses keeps increasing can be determined from the sign sequence formula. When the number of plus reaches $\varepsilon\omega$ then ind x increases by the juxtaposition of a single plus. Since this is precisely $g(c)$ it can be discarded by cofinality. However, consider what happens when the number of plus reaches the first epsilon number above ε. The continuation of the sign sequence in x is precisely ind x. We almost have a generalized epsilon number as discussed in chapter nine. What is missing is that we are not assuming that α absorbs $\varepsilon\omega$ multiplicatively. At any rate, at that point ind x can no longer be discarded, and, in fact, is cofinal by itself as a lower element in the computation of $g(x)$. So $g(x)$ gets an extra plus.

This is similar to what happens in the simple case for ordinals at the first epsilon number. In fact, for any x such that $c \leq x \leq d_\beta$ the same reasoning we used so far including the splitting up into various cases can be applied to the tail of c in x in order to determine the sign sequence of the rest of $g(x)$. Again we are left with only one case to consider further. Specifically, the case consists of an element beginning with c followed by ε_1 pluses for some epsilon number ε_1 greater than ε and then followed by $\varepsilon_1\omega\beta$ minuses for some limit ordinal β.

We now have a set up for a "grand" induction. At every stage we are left with one case to consider further so it appears that we will never be done. The pattern is as follows. We have strings of pluses and minuses, the ith string of pluses has length ε_i where (ε_i) is a strictly increasing sequence of epsilon numbers, and the ith string of minuses has length $\varepsilon_i\omega\alpha_i$ where α_i is a limit ordinal for all i.

What we do next is to consider a sequence b such as the one described above where i runs through the set of all positive integers. By the induction described above we know the value of g for every proper initial segment of b. Now $g(b) = \{\text{ind } b, g(b')\}|\{g(b'')\}$. By cofinality we may limit ourselves to those initial segments b^o of b which consist of full strings. Hence by the inductive hypothesis $g(b^o)$ has the following form: the ith string of pluses has length ε_i, the ith string of minuses has length α_i and a plus is added at the end. (This is valid whether the last string in b^o consists of pluses or minuses.) Also, ind b has the same form but, of course, with no plus at the end. By mutual cofinality we can ignore the plus at the end in $g(b^o)$. This is easy to see since the lower element obtained by stopping after the ith string of minuses then followed by a plus is less than the element obtained by stopping after the $(i+1)$st string of minuses, and the upper element obtained by stopping after the ith string of pluses is greater than the element obtained by stopping after the $(i+1)$st string of minuses then followed by a plus.

Hence $g(b)$ simplifies to $\{\text{ind } b, \text{ind } b'\}|\{\text{ind } b''\}$. If we apply the remark concerning restricting oneself to full strings by cofinality to ind b, it follows that $\{\text{ind } b'\}|\{\text{ind } b''\}$ is precisely ind b.

We have shown that $g(b)$ has the form $\{ind\ b, F\}|\{G\}$ where $F|G = ind\ b$ and F and G are initial segments of $ind\ b$. It follows easily that $g(b)$ is $ind\ b$ followed by a plus. We see this directly as follows. If we denote $ind\ b$ followed by a plus by c then clearly $F < ind\ b < c$. Since G is an initial segment of $ind\ b$ and $ind\ b < G$, $c < G$ by the lexicographical order. Hence $F < c < G$. Since $ind\ b < c$, we see that c satisfies the betweenness condition. On the other hand, suppose $F < x < G$ and $ind\ b < x$. Since $F|G = ind\ b$ it follows that $ind\ b$ is an initial segment of x. Since $ind\ b < x$ it follows that x must begin with $ind\ b$ followed by a plus, i.e. x has c as an initial segment.

It is now easy to see that the situation after the ω strings of pluses and minuses behaves the same way as the situation after ε discussed earlier. Furthermore, the same argument used above for ω strings of pluses and minuses applies to α strings of pluses and minuses where α is any limit ordinal.

Thus our discussion takes care of the value of g for an arbitrary sequence of pluses and minuses. Although the final result could be expressed as a formal theorem we feel that this would make the pattern look more complicated than it is, i.e. we feel that it is most lucid to express the procedure in the somewhat informal manner which we used.

However, we shall end this book by a formal theorem which describes precisely when $g(b) = b$, i.e. $\exp \omega^b = \omega^{\omega^b}$. This result should be of interest since it is a natural culmination of earlier theorems.

Theorem 10.20. $g(b) = b$ if and only if b has either of the following two forms.

(1) b is less than some ordinal $\alpha < \varepsilon_0$ where ε_0 is the first epsilon number and $\frac{n}{\omega} < b$ for all integers n.

(2) b begins with at least ε_0 pluses and the first string in the sequence for b such that the initial segment of b which terminates at the end of the string is not a generalized epsilon number is a string of pluses. Furthermore, if α is the number of pluses then choose ε to be the largest number of pluses such that $\varepsilon \leq \alpha$ and the sequence

obtained by replacing the final string of α pluses by ε pluses is a generalized epsilon number. Then $b > \varepsilon+n$ for all integers n. (Such an ε exists since the l.u.b. of ε numbers is an ε number.)

Proof. The proof is easier than the statement of the theorem!

Essentially it follows directly from the preceding discussion. First suppose that b is less than some ordinal $\alpha < \varepsilon_0$. The case $b \geq 1$ is taken care of by theorem 10.17. The result for $b = \frac{1}{2^n}$ has been mentioned earlier. This is the case of a plus followed by a finite number of minuses. Now the earlier discussion dealing with an epsilon number ε followed by a sequence of minuses applies to a single plus followed by a sequence of minuses, as was also noted earlier. Hence if b begins with a plus followed by a finite number of minuses, g(b) continues the same way as b so $g(b) = b$. Since $g(\frac{1}{\omega}) = 0$, $g(b) \neq b$ for $b \leq \frac{1}{\omega}$.

We still must consider the subcase where b begins with a plus followed by ω minuses, and a plus. The earlier discussion dealing with ε followed by $\varepsilon\omega\alpha$ minuses followed by a plus where α is a non-limit ordinal actually applies to this case. Since we regard ε as 1, $\omega = \varepsilon\omega\alpha$ with $\alpha = 1$. The plus followed by ω minuses contributes the empty sequence to g(b) (we know that $g(\frac{1}{\omega}) = 0$). The next plus contributes a plus followed by ω minuses and from then on g(b) continues the same way as b. Note that we obtain g(b) resembling b for a strange reason: i.e. the plus at the end contributes precisely what has been previously discarded. g(b), however, does not have the extra plus in b. It is thus clear that $g(b) = b$ precisely if the string of pluses following the plus and ω minuses contains at least ω members. This is precisely the condition that $\frac{n}{\omega} < b$ for all integers n.

Now suppose that for no ordinal $\alpha < \varepsilon_0$ does b satisfy $b \leq \alpha$. This is equivalent to the condition that b begins with at least ε_0 pluses.

Now assume that b is not a generalized epsilon number but b consists of c followed by a string of minuses and c is a generalized epsilon number. Then ind b consists of c followed by a string of minuses of length smaller than the corresponding last string in c. Hence ind $b > b$. Since $g(b) >$ ind b it follows that $g(b) \neq b$.

This argument extends immediately to any sequence having b as an initial segment. A very weak use of the sign sequence formula is all that is needed.

If b is a generalized epsilon number then certainly $g(b) > \text{ind } b = b$. We know in fact that $g(b)$ consists of b followed by a plus. Now let b be a generalized epsilon number followed by a string of α pluses. Then $g(b)$ begins with c followed by a string of $1+\alpha$ pluses. Then the situation resembles the one in the proof of theorem 10.17. The distinction between α and $1+\alpha$ is wiped out precisely when α is infinite in which case $g(x) = x$. This completes the proof.

In concluding this book we hope that the reader is convinced that the class of surreal numbers is a fascinating class of objects found in nature. Since the study of these numbers is still at an early stage, we are confident that there are many exciting results which are just waiting to be discovered by an alert reader. For this reason we expect and even hope that much of the presentation here will be improved as new insights are gained.

REFERENCES

[1] Conway, J.H. (1976). On Numbers and Games, Academic Press, London & New York

[2] Knuth, D.E. (1974). Surreal Numbers, Addison-Wesley, Reading, Massachusetts

INDEX